MEDIA AND COMMUNICATIONS - TECHNOLOGIES, POLICIES AND CHALLENGES

TELEVISION AND ROMANCE

STUDIES, OBSERVATIONS AND INTERPRETATIONS

MEDIA AND COMMUNICATIONS - TECHNOLOGIES, POLICIES AND CHALLENGES

Additional books in this series can be found on Nova's website under the Series tab.

Additional e-books in this series can be found on Nova's website under the eBooks tab.

MEDIA AND COMMUNICATIONS - TECHNOLOGIES, POLICIES AND CHALLENGES

TELEVISION AND ROMANCE

STUDIES, OBSERVATIONS AND INTERPRETATIONS

AMIR HETSRONI
EDITOR

nova publishers
New York

NOTICE TO THE READER

The Publisher has taken reasonable care in the preparation of this book, but makes no expressed or implied warranty of any kind and assumes no responsibility for any errors or omissions. No liability is assumed for incidental or consequential damages in connection with or arising out of information contained in this book. The Publisher shall not be liable for any special, consequential, or exemplary damages resulting, in whole or in part, from the readers' use of, or reliance upon, this material. Any parts of this book based on government reports are so indicated and copyright is claimed for those parts to the extent applicable to compilations of such works.

Independent verification should be sought for any data, advice or recommendations contained in this book. In addition, no responsibility is assumed by the publisher for any injury and/or damage to persons or property arising from any methods, products, instructions, ideas or otherwise contained in this publication.

This publication is designed to provide accurate and authoritative information with regard to the subject matter covered herein. It is sold with the clear understanding that the Publisher is not engaged in rendering legal or any other professional services. If legal or any other expert assistance is required, the services of a competent person should be sought. FROM A DECLARATION OF PARTICIPANTS JOINTLY ADOPTED BY A COMMITTEE OF THE AMERICAN BAR ASSOCIATION AND A COMMITTEE OF PUBLISHERS.

Additional color graphics may be available in the e-book version of this book.

Library of Congress Cataloging-in-Publication Data

Names: Hetsroni, Amir editor.
Title: Television and romance : studies, observations and interpretations / editor: Amir Hetsroni, Ariel University, Israel.
Description: Hauppauge, New York : Nova Science Publisher's, Inc., 2016. | Series: Media and communications: technologies, policies and challenges | Includes bibliographical references and index.
Identifiers: LCCN 2016012393 (print) | LCCN 2016029254 (ebook) | ISBN 9781634850773 (hardcover) | ISBN 9781634850926 ()
Subjects: LCSH: Sex on television. | Love on television. | Television viewers--Social aspects.
Classification: LCC PN1992.8.S43 T46 2016 (print) | LCC PN1992.8.S43 (ebook) DDC 791.45/6538--dc23
LC record available at https://lccn.loc.gov/2016012393

Published by Nova Science Publishers, Inc. † New York

Contents

PREFACE

When I started to amass articles for this collection the first question I was asked by colleagues was: "Besides media historians – who is still investigating the impact of television today?" While it is obvious that the landscape of media and consequently of media research is changing with new media becoming a hot topic and conventional media retreating to the back seat, there is also empirical evidence that the impact of new media is considerably augmented when supported by conventional media conduits like television (Jenkins, 2006). For example, studies of recent protest campaigns show that even when initially demonstrators are recruited by social networks like Facebook, a larger number of protesters come when the event is covered by conventional media (Hetsroni & Lowenstein, 2012).

There is no reason not to assume that a similar type of mechanism holds for romance. While scripts of dating behavior and hook-up tactics may first appear in blogs, web sites and social networks, their popularity exponentially increases when they are picked up by successful TV shows. In fact, my very personal story from 2015 demonstrates the capability of television to make a romantic change in one's life even when the original precursor is the internet.

After the general elections that took place in Israel in March 2015 I published a Facebook post about the tendency of poor Jewish immigrants to vote for right-wing parties and added that if we did not open the door to these immigrants – the left wing would have won the elections. The post became viral, garnered criticism and led to my invitation to a popular news magazine in Channel 2 - Israel's most highly watched TV station. There, I was confronted with a second generation immigrant who compared me to Hitler. In return, I told her that "Nothing bad would have happened if your parents had stayed in Morocco and rotted there." This statement had made me the talk of the town – an instant celebrity. Polls have shown that I am "the most famous academic" but also "the most hated person in the country" (Ghert-Zand, 2015). It should be noted that until that point I was merely "the professor who holds somewhat unusual positions on certain issues" (for instance, I wrote once that he state of Israel should have been established in Europe); a cult figure who pops out on the TV screen when a talk show needs an unexpected commentator to complete a panel. But even this previous limelight presence did not prepare me for what was to happen. For weeks, the phone did not stop ringing. People cursed, asked questions, raised concerns and sometimes courted. After some "wild days," wherein I started the night in one bed, switched to another, and ended up in a third bedroom, I chose my new Significant Other - a rich Arab girl, fifteen years younger than me, ten centimeters higher than me, and very good looking. My chances of teaming up with her had I remained a "plain" Jewish academic in a society abundant with a racist conception of intimate relationships would have been close to zero.

While this book does not deal with the influence of television on the romantic life of celebrities, the studies assembled here do shed light on its effect on what ordinary people do in their daily life, romantically wise and on the reflection of intimacy on the small screen. The topics analyzed here include the influence on TV viewing on relationship quality, reflection of elderly love on prime-time programming, effects of romantic films on beliefs about the looks of intimate partners, representation of single men in sitcoms and teen love in drama, and many other topics that are undoubtedly worthy of investigation.

This book is divided into two sections based on the chapters' methodology - qualitative or quantitative. Some may see this division as too rigid or even artificial, but in my opinion qualitative and quantitative research do not only differ methodologically but also represent a different way of looking at the world to an extent that justifies the division.

The qualitative section includes four chapters. The first, by *Colleen Keene*'s entry which analyzes the depiction of sexual desire in two crime series: The Mentalist (American series) and Silent Witness (British program). The reading demonstrates similarities and differences in the construction of romance in British and American crime drama. Keene argues that the British detective drama uses a cool and rather detached tone, whereas the American program is lighter and less severe in its approach. While both shows attempt to engage the audience, the British show works like a novel and the American series employs tactics of popular Hollywood films.

The second chapter, by *Maria-Jose Masanet, Pilar Medina-Bravo* and *Sue Aran-Ramspott*'s, offers a semiotic analysis of forbidden love in current Spanish TV drama aimed at teen viewers. Two successful shows - Los protegidos and El barco – are used as case studies. The analysis indicates that romantic love survives in modern storylines and is still subject to perpetuating age-old myths and gender role stereotypes, despite the requirements of healthy interpersonal relationships in societies of equality.

The third chapter in this section, by *Andrea McClanahan*, reviews the fictional representations of childfree women in five American television shows from the last two decades: *Whitney, The Big Bang Theory, Sex and the City (1998- 2004), How I Met Your Mother* and *Grey's Anatomy*. The author shows how the hegemonic ideal of motherhood is framed in relation to these women. McClanahan distinguishes between two different frames that are used to describe childfree women in popular programming: "romance trumps all," where women who verbally state their desires to not have children are easily swayed when presented with a choice to end a romantic relationship or change their opinion on their stated aspirations in relation to children; and "childfree by choice, romance-free by consequence," which presents women who are committed to being childfree regardless of the consequences. In the second frame, the women are often left without romantic relationships due to their childfree status.

How I Met your Mother is also the focus of the fourth and final chapter in this section, by *Jennifer Whalen*, who uses this program to examine ways in which media representations of interpersonal performances on a prime-time television show speak to and reify the cultural narratives surrounding the stigma of male singledom in contemporary American culture.

The second section of the book features empirical investigations of the representations and effects of romance in television. Five chapters are included here. The first, by *Heather L. Hundley* and *Jo Anna Grant* is a combined content analysis and critical discourse analysis of the portrayal of senior citizens sexual activity in popular US prime time television drama from recent years. This entry sheds light on the romantic behavior of an under-investigated age cohort and shows that senior sex is presented on the screen as an activity that involves single heterosexuals of the same age and from the same race. Interestingly, in contrast with other age cohorts, the elderly show no signs of jealousy, but in consonance with younger protagonists they maintain a wide range of relational commitment - from one-night stands to long term marital relationships. The chapter also indicates that while peers are supportive of senior sexual relationships family members are often not – perhaps due to financial fears. The results are compared with social beliefs and stereotypes of senior sex and with population data regarding senior sex and confirm that television is not exactly a true mirror of older people's life.

The second chapter in this section is *Dorit Zimand-Sheiner* and *Osnat Roth-Cohen*'s content analysis of romance in Israeli television. This entry is different from other chapters in the book in its database – TV advertising. The authors make use of Sternberg and Weis' *triangle of love* model to code a representative sample of advertisements aired by major broadcasters. Only a small minority of the ads presented a plot about a couple's relationship with passion, intimacy and commitment being the most frequent love components depicted in these advertisements. The findings are discussed in the context of the relationships between reality and commercial media. Zimand-Sheiner and Roth-Cohen demonstrate that the presentation of love in advertising has very little connection to the brand because love is constructed in accordance with public perception of romance regardless of the advertised product.

The third chapter in this section is *Amber L. Ferris*' who employs *uses and gratifications* theory to examine viewers' motivations to watch reality dating programming and their feelings about the shows and about dating in general. The author shows that gender, romanticism, voyeurism and perceived realism of the programs predict dating attitudes and liking of the programs. Specifically, women, romantic and pro-voyeuristic persons and people who perceive reality shows as realistic are more prone to watch the programs and like them and to believe in traditional dating stereotypes that reality shows tend to connote.

The fourth chapter, by *Emilio S. Hernandez* and *Narissra Maria Punyanunt-Carter*, is a *cultivation* study. The authors employ this theory, which is more often used to examine the effects of TV viewing, to look at the impact of watching romantic films. They compare the effect of exposure to such films with the effect of watching typical romantic programming and find that the effects are quite similar. Frequent viewers of the romance genre tend to hold stereotypical views about dating and relationship formation e.g., believe in the power of love to abridge personal differences and the capability of beauty to cover for emotional deficiencies. The effect is stronger among females in accordance with the myths typically presented in romance films.

The final chapter of this section and in the book is a summary of my own studies (*Hetsroni*) about the contribution of TV viewing to relationship quality. These are also mainly *cultivation* studies which show that heavy viewing correlates with lower commitment, less satisfaction, greater tendency to engage in conflicts and to seek a pragmatic love style that is devoid of erotic attraction. One explanation of the trend is that the ideal presentation of romance on television leaves the viewers disappointed of what they experience in their own life. A rival explanation is that people who are already disappointed of their love life watch TV as a refuge. This way or another, it is obvious that the way intimate relationships are presented on television is very far from the reality. For instance, while on television couples fight about fidelity – in reality they are more likely to have a conflict about house clean-up.

The causality puzzle posed by my studies remains open. The same is true about the distinction between the relatively small effects detected in the empirical studies that are presented throughout the volume and the subjective estimation that watching television does affect our conception of romance (Galician, 2004). Perhaps, the problem is that the studies examine the ordinary and avoid dealing with the extraordinary. This issue remains open for further explorations. Nonetheless, this collection offers a versatile edifice of research on television and romance. Studies from different cultures and continents that use very different methodologies shed light on the way love is portrayed on the small screen and effect of this portrayal on our conception of romance.

I cannot finish this introduction without thanking those who ought to be thanked: The fourteen authors who joined me and contributed chapters to the collection. I would also like to thank fifteen other anonymous authors whose articles were excluded because they did not pass the review. I am also grateful to my dad, my Significant Other and friends who were supportive while I was struggling with texts, my current institute (Koç University) that helped me find time to complete the work and the people at Nova Science who have exhibited tremendous patience and meticulous consideration during the work. See you all in the next book.

References

Galician, M-L. (2004) Sex, love and romance in the mass media: Analysis and criticism of unrealistic portrayals and their influence. Mahwah, NJ: Lawrence Erlbaum.

Ghert-Zand, R. (2015, March 23). Academic on hot seat after tirade against Mizrahi Jews. Times of Israel. Retrieved from: http://www.timesofisrael.com/academic-on-hot-seat-after-tirade-against-mizrahi-jews/

Hetsroni, A., & Lowenstein, H. (2012). Old media or new media? The influence of media type reliance of social protesters on their level of identification and activism. Connection, 43, 13-22.

Jenkins, H. (2006). Convergence culture: Where old and new media collide. NT: NYU press.

SECTION I: QUALITATIVE EXPLORATIONS

In: Television and Romance
Editor: Amir Hetsroni
ISBN: 978-1-63485-077-3

Chapter 1

UNREQUITED SEXUAL DESIRE IN US AND BRITISH CRIME SERIES: COMPARING THE MENTALIST WITH SILENT WITNESS

Colleen Keene
Independent Scholar

ABSTRACT

Protracted longing and unsatisfied sexual desire are key elements and strategies in romance narratives from Jane Austen onwards. Absorbed over time from novels into film and television scripts, they foster strong audience attachment, with the serialisation mode itself providing the hope for closure and fulfilment that is stretched out further and further. Popular crime television now combines with romance and soap opera to create new genre mixes that have wide cultural appeal. American and British television crime series have features in common, as well as much that separates them. This paper looks at two popular series, *The Mentalist* and *Silent Witness* and the under-stated but ongoing romantic and sexual tension between the two sets of work partners – Jane and Lisbon, and Nikki and Harry. Key scenes in two major episodes are examined in detail to test the romantic conventions and tropes at play. Differences between the approaches, modes and content are explored and some comments are made on national comparisons.

MIXED CODES, MIXED PLEASURES

Genre conventions and codes, whether occurring in novels or film and television, have common effects and appeal. This article analyses episodes of *The Mentalist* and *Silent Witness* in order to tease out the impact of these conventions in two popular current television crime dramas - one from the US and one from the UK. For viewers of television as the popular medium *par excellence*, romance and crime can interact in intriguing ways, presenting strong material for psychoanalytic inflections of content and forces simmering below the narrative surface[1]. Romantic desire is embedded in television viewing, in the conventions, tropes, images, clues, and in symbolic readings of narrative and plot and wider interpretations of meaning. Crime narratives represent a safe vicarious outlet for many difficult or repressed human reactions. With such broad appeal, romance and crime genres have moved from a denigrated place as lower forms in popular culture in contrast to high culture and what is regarded as quality viewing[2]. The emergence of the new stream of high quality, innovative television drama series, starting in the 1990s with HBO and *The Sopranos*, has seen the elevation of new genres and the now widely used term 'the Golden Years of Television'[3].

Crime and detection narratives as a genre have a history dating back to the 19th century and migrated to television in a major way from the 1950s onwards. The field has developed more diverse elements and wider appeal since the 1990s. No longer dominant are the Agatha Christie classics or the genteel mode o*f Midsomer Murders*, and more prominent now is the *Prime Suspect* strand of darker crime and graphic realism, in British series most particularly. American equivalents may be *Law & Order SVU* and *Criminal Minds*. The two series under discussion here are strong examples of contemporary crimes series, *The Mentalist* from USA and *Silent Witness* from the UK. Both employ rich and recognisable genre detail while stretching the forms in new ways. Crime and romance genres central to much popular viewing, intermingle in these two series, enriching both forms.

---

[1] Julia Kristeva, celebrated for her feminist contribution to modern psychoanalytic theory and discussions of 'the Abject,' is also an author of innovative detective fiction, the first title being *Murder in Byzantium* in 2004.

[2] For discussion on romance, melodrama and soap opera in television see numerous discussions in Marcia Landy (ed.) (1991) *Imitations of life: a reader on film and television melodrama.* Michigan: Wayne State University Press.

[3] In the Prologue to *Difficult men: Behind the scenes of a creative revolution* (p.9), Brett Martin (2013) discusses different phases of Golden Age, noting this as the Third Golden Age, with the rise of HBO, serious creative writing and celebrity show-runners.

Mixed genre elements – crime plus romance – account for many of the contemporary viewing pleasures of separate and shared codes and conventions. Melodrama is an older key ingredient in both crime and romance genres, but also has a potent place in the heightened new mix. Romance and soap opera is another of the newer directions, one which encodes emotional and vicarious sexual pleasure for the viewer in many ways, challenging conventions of mode and gender[4] and finding expression in the new genre mixes.

The Mentalist (2008 –2015) is a popular US crime series featuring a team of investigators and a central pair, Patrick Jane, a brilliant consultant and former professional psychic, and Teresa Lisbon, the rather stoical head of the team. Jane has a private emotional burden, the murder of his wife and daughter, which serves as the ongoing and serious individual plot line underlying the ever-changing cases pursued by the team. Over five series Jane relentlessly tracks possible perpetrators, who play deadly games with him. The team serves as his new family.

Silent Witness (1996 –) is a long-standing British forensic crime series, focusing on the team of pathologists Leo, Harry and Nikki (a forensic anthropologist) who work closely with officers of the law to uncover and solve a great variety of crimes, often influenced by social and class factors.

The pairs of partners in the two series – Jane and Lisbon in *The Mentalist* and Nikki and Harry in *Silent Witness* – defy the convention of the solo masculine hero as the key agent in all senses. In general, there is now much more gender and (hetero) sexuality non-conformity and role reversal in television series: both genders are active agents, instigators, subjects and objects, and substantial screen time is devoted to partnerships between male and female protagonists. This is often a heterogeneous mix of professional and personal elements, mostly practical, but sometimes with glimpses of underlying (hetero) sexual attraction. In the case of these two pairings the undertones are distinct and clear: Jane gently pursues Lisbon throughout *TM* and Nikki quietly adores Harry in *SW*.

The scaffold of one-sidedness or imbalance is a convention in itself, and a standard strategy of romance, centralising the prolonged pursuit of fulfilment. Openness and incompletion create intrigue and yearning, and lack of satisfaction sets up a dynamic of viewer engagement and identification. Desire embodied in key characters hooks in the viewer and a sense of anticipation teases out a forward trajectory. Key elements of romance – unrequited desire

[4] See several chapters in Landy (1991), Modleski and Brundson in particular.

and narrative uncertainty – are intrinsic to the medium of television itself, which has long used serialised cliff-hangers to keep viewers watching from week to week or from season to season, propelling the story along. Serialisation lends itself to attachment and anticipation, and the move in crime viewing from discrete episodes to ongoing narrative strands is a feature of the Golden Age that has been called 'novelistic.'[5]

Perceptions and readings of romance have mixed sources and strands: the Romantic literary tradition (poetry mainly) of the 18th century and slightly later in novels - mostly written by women – featuring women's desire and regarded as 'lower' forms'[6]. But the more refined literary novels of Jane Austen and energetic prose of the Brontes also nourished romance as a category and novelistic genre. With regard to television as a medium, these novelistic modes and strands were transmuted in script writing and transmitted into drama. As well, they moved into the broader form of soap opera, a prolific focus since the 1980s. The spectrum covered *Dallas* through to *Thirty Something* in the 80s, running through to *Orange is the New Black* and the more complex mix in current dramas like *Madame Secretary.*

Romance has powerful underpinnings and interpretative sources in developmental theory: the study of human psychological, sexual and emotional growth, with grand foundational contributions from Freud, Klein, and Bowlby. The progress of development is located in different stages and elements. In Freud, models of emotional connection are bound up with parent figures, drives and repression, mostly unconscious – the search for completion and the playing out of unresolved anxieties of the under-developed self. Psychoanalysis, especially the feminist stream (e.g., Helene Cixous, Luce Irigaray, Laura Mulvey[7]) has also focused on pleasure and the maternal sources of (pre-Oedipal) pleasure. Klein centres on the breast and place of the love object in the growth of relationships and the 'splitting,' projection and introjection (of primary figures and experiences). Bowlby explores the 'secure base' of self and unhealthy working models and forms of attachment. All focus on the move from primary, originary relations in the development of connections and attachments. All focus on ways in which connection can be inadequate and unsuccessful, go awry or turn into pathology. The genesis of

---

[5] See Toby Miller (ed.) (2003). *Television: Critical concepts in media and cultural studies.* London: Routledge.

[6] See major discussion in Radway's *Reading the romance* – first published in 1984, revised edition 2007. Chapel Hill, NC: University of North Carolina Press.

[7] See *New French Feminisms* (1984) Elaine Marks and Isabelle de Courtivron, for Cixous' *jouissance* and Irigaray's 'feminine.' See Mulvey's discussion of the male gaze in *Screen 16* (1975), 6–18.

blocks and arrested development often focuses on prolonged unrequited desire – are they about repressed attraction, wish fulfilment or attachment disorder? Or are they about projected unmet need and repeated primary experiences (of neglect, abuse, intrusiveness) from the family of origin[8]? Unrequited desire, in psychological terms, appears pathological in origin; something unresolved from early experience, often resulting in the fear of intimacy. It is insecurity and the fear of intimacy and loss that seem to underpin the representations under discussion here.

The persistence of under-stated, unexpressed sexual desire across time – in this case across many seasons of *The Mentalist* and *Silent Witness* – deserves some speculative psychological reading. And delayed gratification is an excellent strategy for writers to keep viewers attached and wondering, desiring viewer fulfilment and satisfaction of a heightened romantic kind. The backstories of characters add to this effect. Patrick Jane's imminent capture of Red John after six seasons of the chase points ahead to his likely separation from Teresa Lisbon. Nikki Alexander's complex relationship with her father and his death feature in the episode prior to the one under discussion, and may heighten the emotional ground for Harry Cunningham, attentive to her in intimate ways at the funeral. But it is the dynamics of the codes, scene structure, images and moods that matter; the elements constructing and working up moments of heightened connection in the episodes and scenes per se.

The Mentalist - Patrick Jane's Romantic Game with Teresa Lisbon

Like many recent crime series, *The Mentalist* mixes its elements while also subscribing to the classic conventions of crime and detection. One of the developments in the crime genre includes the employment by the police force of gifted amateurs (e.g., *Medium, Person of Interest*) who assist the police and provide other dimensions of interest and action. The twist of *The Mentalist* is that Patrick Jane (Simon Baker) is a once-renowned psychic who now admits to simply having deep skills of perception and intuition. But he also has his personal quest: to catch the killer of his wife and child, which generates the time shifts and flashbacks across the series. Over time, and over five series of

[8] See Ride's *Echo and Narcissus* (2002). The main term for this phenomenon, developed in Object Relations theory, is 'projective identification.'

the show, Patrick Jane has slowly but surely isolated five varied and distinct suspects in the long-term case he is personally pursuing amid the wide array of cases that he works on with the CBI team headed by Teresa Lisbon (Robin Tunney). And Jane is a lovable rogue, prone to pranks and teasing, using all sorts of unconventional means to assist the CBI (California Bureau of Investigation) cases and operations. Clever tactics, facetiousness and undercurrents of attraction have featured explicitly in their partnership for over five seasons.

Series 6, Episode 6, 2013: 'Fire and Brimstone'

I will analyse the sixth episode from the sixth season. At the juncture of this episode, Jane approaches each suspect and lures them to a meeting at the former Jane family residence at Malibu with the promise of information and final exposure of the identity of Red John – the serial killer and murderer of his wife and child. Typically inventive, Jane induces and threatens the five men to come to a cabin on the property, and the episode opens as he sets up the weaponry in the shadowy gothic setting, awaiting his prey. The stated intention is to expose and confront Red John. The scene shifts back two days as the team discusses the *modus operandi* and plot complications surrounding the various suspects. Lisbon, in loyalty to their long and intense partnership, insists on being there when he "reveals" Red John as her condition of approval. Typical of their toing and froing, Jane hedges, disobeying orders being part of his *modus operandi*.

Later in the episode, in their next specific conversation, at the house and amid shifts in time and setting, Lisbon sees the gun and knows he will not stop at arresting Red John. She persists and, to his shock, indicates that she is prepared to violate 20 years of professional behaviour to accompany him on his dubious illegal take-down and won't stop him. "Some men don't deserve a trial or a jury. They deserve what they have coming to them" she argues. Jane is worth the physical and ethical risk. They have worked together for many years and she isn't about to back out now. They are steadfast partners, and much more. But is this just more seductive verbal play between them? Jane seems to agree to her presence but, ever attentive to her safety and reputation, is highly reluctant to allow her to be involved in the confrontation that is likely to end in the assassination of Red John. Jane, ever resourceful and ever gallant, foils her attempt to be present and leaves her stranded by the roadside while he takes off in her car to get to the scene and follow his dangerous take-down alone – but it is the nature and means of his foil, his bluff, that are most interesting. The trick is a trick on one level only, the practical level. Otherwise it is sincere, a declaration emerging at a pivotal moment, in more ways than one. And the scene hinges on romantic tropes.

Patrick Jane and Teresa Lisbon, the senior CBI detective and the consultant with extraordinary (psychic) powers of intuition and perception have been circling each other for five series. Playfulness and jokey sarcasm are their mode of engagement, with Patrick at times causing Lisbon great frustration with his antics, trickery and evasiveness. The elusiveness and play is an often comic narrative distraction that diverts, but never quite blocks, the tight plot flow and also has a distinct and potent edge of romance: the advance / retreat of romantic sexual tension. The deferral of satisfaction and the erotic charge of waiting and delayed gratification are classic strategies of romantic drama from Jane Austen onwards.

Patrick is brilliant, charming, handsome and flirtatious. Lisbon is attractive in her competence and coolness. In themselves these are coded romantic norms and stereotypes. The pursuit is quiet and understated but becomes more pronounced as the series progresses and we, the viewers, see Lisbon's unfolding engagement and the deeper attachment growing between the pair. It is a cleverly constructed slow burn of unfolding desire and this pivotal episode, important as well to the overall Red John plot, concentrates and spells it out in a more obvious way. Patrick announces his feelings for Lisbon in a new way, but then literally retreats, leaving Lisbon and the viewer in a state of arousal and expectation.

Late in the episode, in an unusual shift, the focus briefly narrows to the pair. Jane tricks Lisbon into getting out of the car (on the way to 'Malibu Bluff') during the pensive drive to see the sunset and share something with her that he has been waiting too long to say – classic tropes of romantic strategy. He announces how important she is to him. "I need to say this now…You have no idea what you have meant, and mean, to me" he says, and they hug strongly, hinting at the likely consequences of his action and their possible separation. She is intensely awkward to hear his words, but deeply moved as well. He then says he needs to get a surprise, something for her from the car – tantalisingly – setting up the expectation, if not of a ring then of some token of affection. Instead he approaches the car – her car – jumps in and drives off, leaving her stranded, without her cell phone. Symbolically this renews the delay all over again, to be traced through many more episodes yet, as the audience sense of anticipation continues to grow alongside the quiet romance.

It is also a form of indulgent frustration of desire and in the manipulative nature of soap opera to tease and withdraw in a literal delaying of climax.

After his shock exit, the also resourceful Lisbon hijacks a car and its driver's phone and rings Jane, pleading: "Don't do this, not without me… Do this and you're throwing your life away." Subtext is 'their' potential desired life together. Jane is adamant "Bye Lisbon," he says, and cuts her off. Jane visits the scene of the original crime, full of quiet emotion. Then he coolly arms, greets and interrogates the five suspects and the unfolding plot of the Blake Association, a corrupt internal ring in the force. Lisbon calls for back-up but gets there first. The suspect with a three-dot tattoo is presumably Red John, but Jane finds that three of them possess it, and it is the marker of involvement in the illicit activities of the association. He singles out the three.

A huge explosion and ensuing destruction of the cabin as Lisbon runs towards it marks the climax, and end, of the episode. The camera suspends focus on Lisbon in full frame – direct witness to the explosion. Her distraught face closes the episode. The final image is then a portrait of loving concern. The awful prospect/possibility of losing Jane lies beyond clear-cut professional engagement, beyond any possible professional ramification of crimes under investigation and missteps in the process.

This episode prefigures the narrative arch of separation, and much more delay over many episodes and locations, with Jane hiding out in the tropics after killing the actual Red John. The pair, and the audience, wait with protracted longing, for reunion, reconciliation and consummation.

Silent Witness - Nikki, Harry and Quiet Passion

Silent Witness is, by contrast, more stereotypically British. It is more formal in structure, realist in treatment and restrained in tone; and more literary, as such. Yet it is also more visceral, graphic and gruesome than *The Mentalist*, in keeping with earlier British crime drama, exemplified by *Cracker* in the 1990s. As well, it is an even slower romantic burn with very few moments of overt desire and little hope of gratification for viewers; a very slow tease certainly, but within the context of grittier, more harshly realistic series focused on the detailed settings of violent urban crime and life of the forensic team.

As with *The Mentalist,* Nikki Alexander (Emilia Fox) and Harry Cunningham (Tom Ward) are professional partners in a crime series, but this one has become a forensic classic, in the more typically American canon of forensic crime dramas, dating back to *Quincy ME* in the 1970s and culminating with the CSI franchise. These dramas focus intensely on the minutiae of detection, of generally scientific and physical detail. This in itself becomes a form of erotic sharing between the partnering forensic pathologists

Harry and Nikki plainly work well together and like each other, but their contact is limited to the lab most of the time. Harry's relationships feature more in the series than do Nikki's; he is presented as the sexually active one, while Nikki waits on the sidelines and watches. The female gaze is more evident in this series, as Nikki's lingering gaze on her partner sustains the viewer's romantic objectification of the handsome Harry. The viewers are never quite sure of Harry's interest or desire until very late in the series (but their sexual liaison is brief).

As in *The Mentalist,* the advance / retreat of the dance of unrequited desire between the two competent colleagues is evident, but is far more under-stated and less central, glimpsed and visited occasionally rather than overted in every episode. Competence and shared work becomes a mode of romantic communication in itself. Pieces of uncovered knowledge often stand in for personal and emotional exchange, with a sexual charge, detail by loving detail. This is different from the broad sweep in *The Mentalist,* a series with multiple cases occurring around an ongoing thread and driving plot line. *Silent Witness*, in contrast to *The Mentalist,* moves episode by episode without the sustaining drive towards a narrative climax.

Series 15, Episode 2, 2012: 'and Then I Fell in Love' - Part 1

I have chosen to analyse the second episode of series 15 of *Silent Witness* because in this episode the focus shifts from the professional laboratory to the domestic and intimate. With the challenge of noisy building works at Harry's apartment block his sleep is badly affected and he tries, unsuccessfully, to nap at work. Later, as he tries to unwind, relax and luxuriate in the bath – giving the viewer a very different angle on Doctor Cunningham – an explosion blows the bathroom apart, momentarily creating viewer anxiety and concern. The gas explosion means he cannot stay where he lives any longer. In a moment of dry humour he tells his partner "My flat blew up," and the ever-attentive Nikki invites him to stay at her home for the duration, while his home is under repair. "You can stay at mine…" she says. "But it's not right" he replies, both facetiously and seductively. She dangles her keys in front of him, and giggles, pre-figuring the sexy tension at night between the adjoining bedrooms at her place. "Ready to go?," "Yes, honey" they play-act – and play house – deciding to have takeaway. Nikki remembers the flowers he left and Harry replies, "No need to say thank for a thank you" as they exchange frankly loving looks. The normally understated attraction is in sharp focus in this episode.

This episode is more complex than the interaction of the team and its case detail suggest, and draws on serious contemporary social concerns of race, class and culture. Romance and its antitheses – anti-romance – provide the outer framework of the erotic moment between the central pair. The episode – titled `And Then I Fell in Love' – is layered and, as usual, contains two cases, but the main trajectory surrounds school girls, young women between 15 and 18 years-old, and a group of angry alienated south Asian young Muslim men ("Pakis") running a cab service. The men brutally groom them ("white whores") for prostitution, using seduction and consumer items, buying them sexy outfits and appealing to their sense of romance; the romantic allure of exotic men and illicit adult sex. For these men, however, it is punishment of the hollow West, its lack of finer culture and its slutty women. In a shockingly self-serving twist, they spout morality while using the girls in the worst form of opportunistic exploitation.

Intimate interiors – beds within bedrooms (and hospitals) are featured throughout the episode which alternates between the soft interior of Nikki's home and the harsh apartment / prison the young men have set up for the purpose of prostituting the young girls. Unrequited desire is juxtaposed with rape. Nikki and Harry sleep innocently, or lie awake, aroused by the other's proximity. The character Amy lies on a red satin bed doped into submission by the charismatic leader Umar, receiving men in a haze of pain and physical trauma. The juxtaposition of intimacies lends a complexity that offers more personal dimensions and stark social comment than in *The Mentalist*.

At the beginning of this episode, as the young men hang out at the edges of the urban scene, Nikki witnesses a young woman, Shannon Kelly, hit by a car, as she flees from somewhere in distress. Nikki visits her in hospital, forms a connection and tries to uncover what is at the heart of her trauma. Her stolen keys emerge when Shannon also flees the hospital and turns up at the house to take the refuge offered by contact with Nikki. The sounds of the break-in bring the sleepy pair, Nikki and Harry, out into the hallway in their pajamas. The stark and violent abuse of the distraught girl throws the gently erotic interaction between Nikki and Harry into high relief.

The episode title – `And Then I Fell in Love' – is tantalising, but does not refer to Harry or Nikki but rather to Shannon, who later admits to Nikki that she fell in love with the brutal ring-leader, Umar. Younis, the sidelined overweight and uncool member of the gang who barely belongs, is the only one with a conscience or sense of honour or empathy. He falls for Hannah, one of the next round of girls being groomed ("I think you're pretty"), but he is cast as the loser with unappealing looks, and has no chance with her, though she senses his care. Any form of care is anathema to the others, and Younis puts his life on the line to save Hannah in Part 2. This is another romantic frame cutting across the anti-romance of the brutal seduction and prostitution racket.

In Part 2 Nikki and Harry dance around each other, in the hallway and at work, discussing the alarming, but also comic, intimacy of sharing head lice from the victims. But the case is harsh and crimes shocking: "We're all weak" announces Nikki in reply to Harry's comment about the father's panicked response in the other case-focus of the episode, hiding his daughter's body in the suitcase. At the end – after Hannah visits the dead Younis on the slab and is ready to name the perpetrators, and Shannon's south Asian step-father avenges her suicide by luring Umar to a violent death – Harry is returning to his own space. "I like having you here" Nikki confesses. "You know…" he explains to her, joking drily: "familiarity … contempt." It takes Nikki longer to get settle back into her ordinary life and get used to his absence: "Harry, is that your phone?" she calls from her bedroom as the last stage of the case unfolds, but he is gone. And their enduring romantic dance of advance/ retreat slows right down again.

Another possible reading of all this unrequited desire is *masochism.* Is there a combination of both pleasure and pain, or pleasurable self-punishment, involved in unrequited, unconsummated, often one-sided attraction and desire? Psychoanalytic readings try to locate the most sexual-emotional patterns and dynamics in early experience. In classical Freudian terms, masochism is a feminine trait and experience of passivity in the face of cruel (father-related) behaviour – a deviant source of pleasure in pain. Julia Kristeva's theory of desire and Abjection are also relevant here [9]. Masochism has been the focus of much cultural representation in recent times, as well as the dynamics of sado-masochism – for example, film representations *Last Tango in Paris, The Night Porter, 91/2 weeks, Fifty Shades* and episodes of several dramatic programs from the past decade like *ER, House* and *Six Feet Under.* Yet, in these pairs – Jane and Lisbon, Nikki and Harry – the dynamics are more complex, subtle and emotional and less overtly sexual. Are Lisbon and Harry sadistic in their

[9] See Kristeva, J. (1982). *Powers of horror: An essay on abjection ideas.* New York: Columbia State Uni.

games and diversions and denials of actual expression? Are Jane and Nikki masochistic in their unwavering attachments to their love objects and undying hopes of satisfaction, despite evidence that is mostly patchy, evasive and unreliable? The viewer is implicated in the sado-masochistic game by also attaching, and remaining attached or addicted, to the playing out of romantic strategies in popular television series.

Concluding … or Not, As the Case May Be

The erotics of waiting and the tension of delayed gratification are intrinsic to the pleasures of viewing television series, whether hedging and edging towards denouement and narrative closure or for climaxes of other kinds. Both these episodes of *The Mentalist* and *Silent Witness* have complex architectural structures and mixed and shifting narrative threads and themes. Both formats and examples at once highlight, exploit and circumscribe the pair bonding in the episodes. Both share the erotics of under-stated, protracted connection and waiting and the frustration of unfulfilled desire. And both tease the characters, and the viewers, in ways that are typical of soap opera and romance traditions.

Both episodes demonstrate the workings of crime and romance codes and tropes, but with some distinctive differences that emerge from different national and cultural traditions. British detective and crime fiction allegedly began with Wilkie Collins melodramatic *The Moonstone* (1860) but Arthur Conan Doyle's Sherlock Holmes and Agatha Christie's Miss Marple and Poirot are its classic exponents, with evidence becoming the core of the `scientific' crime method and mode. With its cool, more detached mode *Silent Witness* has a specific intellectual appeal, in keeping with the literariness of UK crime drama.

The engagement is more broad and populist in *The Mentalist*, less formal and literary. Raymond Chandler and Dashiell Hammet are the founding figures of US crime fiction, both major stylists and hugely popular. But television crime genre in the US is conventionally more energetic and less low-key and cool, right from the days of classic high-action drama like *Hawaii Five-O* and *MacGyver*. The new wave of US forensic series includes shows such as *CSI, Criminal Minds* and *Bones* that have more in common with *Silent Witness,* but they too are less focused on the social order and social realism than the narrative intrigues of evidence. And the high-quality soap opera elements of *Hill Street Blues* would seem to contribute more to the lighter atmospherics of *The Mentalist* than the violent edginess of *The Shield.*

Arguably, TV series embody aspects of the cultural traditions and leanings of their respective nations. In our case, both *The Mentalist* and *Silent Witness* celebrate the attractions of the crime genre, but with different emphases. *Silent Witness* evokes the interest in the dark and violent tradition of Jack the Ripper and other true crime cases, with markedly more interest in contemporary social concerns and class context. Social realism is a hallmark of *Silent Witness* and most British drama since the 1950s, with class awareness as the broad frame. *The Mentalist* evokes the classic PI tradition in some ways, such as Patrick Jane's acuity, but within a more generalised and broader commercial mode and with a lighter touch. The romantic genre elements of the US series are more broadly popular and have more links to melodrama and soap opera, while the British series is closer to the formalism of literary romance.

The romantic elements and strategies *within* successful, high-rating television series enact the romance of the medium itself and its universal appeal – vicarious experience, second-hand pleasure, love objects, anticipation, objectification, delayed gratification, addictive potential – all intrinsic ingredients and pleasures of the serialised viewing experience. The hybridisation of crime and romance in television intensifies the addictive qualities of television. The emotional appeal and power of romantic drama (including melodrama and soap opera elements) substantially enrich the crime genre and expand the audience and reach of contemporary viewing culture and experience.

REFERENCES

The Mentalist (2008 – 2015) Series 6, episode 6, 2013: 'Fire and Brimstone.'

Silent Witness (1996 –) Series 15, episode 2, 2012: 'And Then I Fell in Love.'

Allen, R.C. (1991). A reader-oriented poetics of the soap opera. In M. Landy (ed.), *Imitations of life: a reader on film & television melodrama* (pp. 496–524). Detroit, MI: Wayne State University Press.

Betchen, S. J. (2013). Introduction: Understanding the pursuer – distancer dynamic. In *Intrusive partners – elusive mates: The pursuer-distancer dynamic in couples*. London: Routledge. [Object relations]

Brundson, C. (1991). *Crossroads:* notes on soap opera' In M. Landy (ed.), *Imitations of life: a reader on film & television melodrama* (pp.466–472). Detroit, MI: Wayne State University Press.

Casey, B. et al. (2008). *Television studies: The key concepts.* (2nd ed.) London: Routledge.

Cooke, L. & Creeber, G. (2008). Drama: the crime series. In G. Creeber (ed.), T. Miller & J. Tulloch (assoc.eds.), *The television genre book* (pp.29–34). London: British Film Institute/Palgrave Macmillan.

Goodall, J. (2015). 'Out of the Comfort Zone,' review of BBC's *Line of Duty* and Helen Piper's *The TV Detective: Voices of dissent in contemporary television* review. Retrieved 1 August 2015 http://insidestory.org.au/out-of-the-comfort-zone.

Holmes, J. (1996). From attachment to intimacy. In *Attachment, intimacy, autonomy: Using attachment theory in adult psychotherapy* (pp.3–34). New Jersey: Jason Aronson.

Kristeva, J. (1982). *Powers of horror: an essay on abjection.* New York: Columbia University Press.

Landy, M. (ed.) (1991). *Imitations of life: a reader on film & television melodrama*. Detroit, MI: Wayne State University Press.

Lembo, R. (2007). Components of a viewing culture. In H. Newcomb (ed.), *Television: The critical view.* (7th ed.) (pp. 455–470). London: Open University Press.

Martin, B. (2013). *Difficult men: Behind the scenes of a creative revolution.* New York: Penguin.

Miller, T. (ed.) (2003). *Television: Critical concepts in media and cultural studies.* London: Routledge.

Modleski, T. (1991). The search for tomorrow in today's soap operas. In M. Landy (ed.), *Imitations of life: A reader on film & television melodrama,* (pp.446–465). Detroit, MI: Wayne State University Press.

Mulvey, L. (1975). Visual pleasure and narrative cinema. *Screen, 16*, 6–18.

Newcomb, H. (2005). Reflections on TV: the most popular art.' In G.R. Egerton & B.G Rose (eds.), *Thinking outside the box: A contemporary television genre reader* (pp.17–36). Lexington, KY: University of Kentucky Press.

Radway, J. A. (2007). *Reading the romance : Women, patriarchy and popular literature*. [2nd ed.]. Chapel Hill, NC: University of North Carolina Press.

Ride, Carol (2002). *Narcissus and Echo: Reflections on narcissistic themes in couple relationships.* Vic., Australia: Psychoz Publications.

Seiter, E.& Wilson, M.J. (2005). Soap opera survival tactics. In G.R.Egerton & B. G. Rose (eds.), *Thinking outside the box: A contemporary television genre reader* (pp.136–155). Lexington, KY: University of Kentucky Press.

Thorburn, D. (2007). Television melodrama. In H. Newcomb (ed). *Television: The critical view.* (7th ed.) (pp.438–450). London: Open University Press.

Turnbull, S. (2010). Crime as entertainment: the case of the TV crime drama. *Continuum: Journal of Media & Cultural Studies, 24*(6), 819–827.

Turner, G. (2008). Genre, hybridity and mutation. In G. Creeber (ed.), T.Miller & J. Tulloch (assoc.eds), *The television genre book* (pp.8–9). London: British Film Institute/Palgrave Macmillan.

In: Television and Romance
Editor: Amir Hetsroni
ISBN: 978-1-63485-077-3

Chapter 2

THE SURVIVAL OF THE FORBIDDEN LOVE IN TELEVISION FICTION: "ROMEO AND JULIET" IN CONTEMPORARY SPANISH SERIES FOR YOUTH

Maria-Jose Masanet[1], Pilar Medina-Bravo[1] and Sue Aran-Ramspott[2]
[1]Universitat Pompeu Fabra UPF, Barcelona, Spain
[2]Universitat Ramon Llull URL, Barcelona, Spain

ABSTRACT

This chapter is concerned with the study of television fiction targeting young people and produced in Spain. The aim is to detect the romantic love stereotypes that may be present in these series and the way they are expressed. It uses as a case study two highly popular teen drama series that were first shown in 2011: *Los protegidos* and *El barco*. The narrative analysis indicates that romantic love survives in modern storylines and is still subject to perpetuating age-old myths and gender stereotypes as a dominant masculinity and a submissive femininity despite the requirements of healthy interpersonal relationships in societies of equality.

> Being without Culebra is like living in the dark ... and I will not live like this
>
> Sandra, Los Protegidos

Erikson (1972) spoke of the media as manufacturers of "stereotypes, often forcing young people to live out caricatures of images that they had initially projected experimentally" (Erikson 1972, p. 161). Television provides teenagers with information not only about the world around them but also about what is expected of them. Through television images, teenagers are able to take a tour around different models and social alternatives and "decide" on the identity that they think might afford them security and appraisal within their group of reference. Furthermore, the audiovisual text is also an element of comprehension of social reality, thus becoming a reference point for information on everything around them (what adults do, what adult life is like, what type of conflicts adults experience, etc.). The audiovisual text will eventually become a cultural text that will become implanted in real experiences, to then become incorporated as a reality itself. Although the teenager may be fully aware that it is fiction and is not to be confused with the real world, numerous studies show that teenagers are not only driven by entertainment when they watch series, but equally by the "inferred" information they think they are being provided on adult and adolescent life (Vives 2007).

> Being young means acting and positioning oneself in the way that the media show us that young people think, feel and behave. This identification with teenage models reinforces the experience of normality within them... The media do not only say what young people are, they also contribute to configuring and defining the forms of the teenager's social identity" (Von Feilitzen 2004, p. 9).

Adolescence is a time that portrays different endings such as that of the body and childhood innocence. Yet it is also a time for young people to initiate new possibilities and new responsibilities: physical changes, a greater capacity to think about themselves and the world around them, discovering sexuality, sexual attraction and the romantic relationship that becomes a focal point of interest, etc. Therefore, understanding the importance of sexuality and the romantic relationship in adolescence requires approaching the emotional, social and cultural complexity involved. As stated by Erikson (1980), loving and wanting to be loved is also part of the construction of teenage identity. At this stage, the awakening of sexuality is linked to the changes in the body itself, the discovery of the other as an object of attraction, curiosity about both these physical and emotional changes, falling in love and initial attempts at emotional intimacy. Sexuality and falling in love therefore become key issues in adolescence.

For this reason it is important that adolescents receive an education that provides them with the resources to be able to cope at this individual time of change. Formal education, in this sense, has instigated attempts at training adolescents in sexuality and love relationships but in many cases this formal education has been polluted by a partial and insufficient focus on love and sexual relationships, as various authors indicate (Bay-Cheng 2003; Kirby 2002). Furthermore, for teenagers it is paramount to be well-informed in these areas but there must be something lacking in this formal education when there is still a high rate of risky behaviour. Meanwhile, the media have gradually been conferred an increasingly important role as informers of sexuality and love relationships, to the point of becoming socializing agents in these issues (Ward 1995; Hetsroni 2008), and might be considered informal educational agents. In fact, diverse research (such as Masanet and Buckingham 2015; Pindado 2005) suggests that the media are an important source of informal education about sexual and love relationships. Investigations in this regard have proliferated in recent years and an increasing number of teenagers consider the media to be a more useful source than school or parents for learning about love relationships and sex (Buckingham and Bragg 2004). Yet it seems tremendous panic still arises from the representations the media may make about love relationships and sexuality. However, we cannot deny these informal educational practices as "students already learn from the media, and they appreciate not only the information they can glean but also the media's appeal to them as knowledgeable and mature" (Bragg 2006, p. 329). Adolescents not only watch the love and sexual relationships the media present, they also discuss them with their peers, friends, family members… and, in one way or another, they end up becoming points of reference. This means the media have become an informal source of sexual and romantic education that teenagers both appreciate and with which they feel comfortable. More specifically, teen series take on a leading role in this sense to hold peak positions in young people's media preferences (Fedele 2011).

As previous research has determined (Coltrane and Messineo 2000; Davis 1990; Fejes 1992), media portrayals of the members of a couple are clearly different: the identification with woman characters is based on their emotional and dependent necessities; men characters, in contrast, tend to be shown as independent and powerful. Love relationship media models have an influence through the viewers' identification or, for the most part, through recognition of the common social universe (Galician 2004). In youth audiences these identification or recognition mechanisms play a vital role as active elements in the process of constructing a personal identity (Fernández-Cavia 2002; Huertas and França 2001) and also as indicators of what young people expect from a love relationship. Is Spain, previous studies (Aran, Medina and Rodrigo 2014) show how romantic love survives, particularly in the early stages of a relationship, and that it is produced from a demand for individual freedom. However, the remaining forms of relationship fluctuate between young people's expression of the desire to renew roles and the deep-rootedness of a social (and affection) model based on a dominant masculinity and a submissive femininity. These studies alert us to the need for research to analyse media interpretations of love and sexual relationships on television. This work falls within this area and, more specifically, on the study of intimacy in fictional series targeted at young people and produced in Spain.

The Myths of Romantic Love in Television Fiction

It must be highlighted that the representation of romanticism has enjoyed certain predominance in fictional series. As indicated by Moran (2003, p.11) "the most significant role that television plays is the reinforcement of existing perceptions about romantic relationships." Romanticism that is linked closely to different myths such as "the power of love," "predestined true love," "love as a possession" and "the total gift of love" may, in fact, become highly impoverishing and imply gender differences within the love relationship.

Several studies have underlined the predominance of romantic love representation in television series. American studies analyzed shows that reject the hegemony of romantic love representation with gender differences and submit new perspectives on gender constructions and love relationships. Works by Banks (2004) on *Roswell* and *Smallville* and Bavidge (2004) and Osgerby (2004) on *Vampire Slayer* demonstrated how these shows give fresh constructions of heroes and heroines in adolescent melodramas that manage to break with certain gender stereotypes on the roles of boys and girls in the field of emotions inside and outside relationships. However, as they also indicate, many of the dynamics represented previously in the field of love relationships still perpetuate. Hills (2004) looked at the series *Dawson's Creek* and demonstrated how the program presents a combination of traditional and post-modern relationship models that confer the series a certain quality. In the words of the author: "by depicting different types of reflexive love relationship, and combining aspects of traditional and pure relationships, Dawson's Creek bids for quality status through its textual and romantic sophistication" (Hills 2004, p. 57). Likewise, other research has focused on series that devote a significant space within their storylines to somewhat neglected issues such as homosexuality (Barker and Andre 1996; Davis 2004) or break with stereotypes in their sexual representation (Douglas and William 2004; Masanet, Medina and Ferrés 2012).

Meanwhile in Spain, Medina and Rodrigo (2009) analysed the Spanish series *Los Serrano* and concluded that it appeals to gender stereotypes within love relationships where romanticism is associated with women through myths such as the existence of the other half, the idea that a woman can change a man or improve him, associating erotic-sexual passion with men and romantic passion with women, presenting men that are educated by parents to be seen as sexually powerful in order to make plain their masculinity and showing jealousy as an indicator of love, while girls are presented as affirming their right to sexual pleasure but always as a pleasure subject to romanticism, a sphere that is affirmed as inherently feminine. Luzón Ramos, Recio and de la Peña (2011) also illustrate some of these gender stereotypes, drawing attention to the fact that fiction represents boys' first love while for girls it is breakups and jealousy. This study suggests that impossible relationships and the myth of sacrifice within relationships are conferred a specific space within television series. Capdevila, Tortajada and Araüna (2011) and Capdevila, Crescenzi and Araüna (2013) analyse the series *Sin tetas no hay paraíso* and underline that the relationship presented by the main characters in the series is presented as a "true love" that is associated to various ideals such as what unique love is, love at first sight, the inevitable, but consistently, that it can overcome all obstacles, etc., thus playing out the traditional myths of romantic love. Meanwhile, Simelio (2010) analyzed a series created specifically for the Internet -*Qué vida más triste* and *Becari@s*-, based on the idea that they might offer fresh content that is more entertaining and varied and concludes that in the end they put forward the same stereotypes as general series in terms of how relationships are portrayed: "demeaning stereotypes are displayed in relationships, along with the focus of women as a source of unhappiness for men and as impossible to understand. They are also represented as weak, capricious and easy to manipulate" (Simelio 2010, p. 7).

In what way do these Spanish programs differ from American counterparts? Guarinos (2009) compared North American and Spanish series and found that American shows sell an image of innocence and morality that camouflage numerous sexist, classist and racist stereotypes while Spanish series present teens with less idealized lifestyles and demonstrate a more risky or tough love and sexual relationships, with looser morals: "in our series sex is explicit, including oral sex, there are homosexual characters, students that have relationships with their step-fathers…" (Guarinos 2009, p. 210). Crespo (2005) echoes this when highlighting the differences between North American and Spanish representations. However, despite the fact that several studies indicate the risky proposal of some Spanish fictional series compared to American series, in the end, Spanish production still represents certain romantic love stereotypes, as indicated in previous studies and as we will see in this analysis.

OBJECTIVES OF STUDY

We will consider the myths that prevail in the development of romantic storylines of dramatic series targeted at teenagers and produced in Spain. The aim is to detect the romantic love stereotypes that may be present in these shows and the way they are expressed and produced and point at "meta-myths" that emerge from the analysis.

In order to carry out the research, a sample of two Spanish fictional series with the highest ratings were selected: *Los Protegidos* and *El Barco*. Both are Spanish productions that were broadcast by *Antena 3*[1] during the prime-time hours for three seasons (2011-2014). Both shows achieved high ratings in Spanish television listings, especially in their first two seasons, and were characterized by the large number of teenage characters. This could have contributed to young viewers' identification with the characters. The two series are included in the category of 'teen series,' according to the criteria established by García-Muñoz and Fedele (2011): they are aimed at a young audience, their episodes last around 60 minutes, they focus on telling the story of young people or teens, mainly during the high school period, and most of the plots are about interpersonal relationships –especially love and friendship-.

Los Protegidos is a series produced by *Boomerang TV* and aired by *Antena 3* with great success. The series falls within the fantastic and drama

[1] Antena 3 is a private Spanish television channel that started broadcasting in 1990. This channel has been broadcasting a variety of Spanish produced series aimed at teenagers and young people such as Física o Química, El Internado and Pulseras Rojas.

genre aimed at all members of the family. It is an ensemble series with characters of different ages. A large part of the main storyline is centred on young people, which is what enables the product to target this age group, and therefore, be suitable for the study. The series' plot is based around the story of a group of children and teenagers with special powers, the father of one of them and a woman who is looking for her kidnapped daughter who all pretend to be one family. They are forced to live like this in order to flee from the *Clan del Elefante*, an association devoted to kidnapping children and teenagers with powers to use them for their own benefit. The aim of this new family is to discover why they have special powers and find Blanca, the kidnapped girl. Meanwhile, they have to attempt to live as a real family, hiding their secret from everyone else.

El Barco was produced by *Globomedia* and topped the ratings chart in Spain. Some first season episodes drew more than four million viewers. It is an ensemble teen drama series with characters of different ages –children, teenagers/young people and adults–. However, most of the characters are teenagers or young people as the series tells the story of a school boat that sets sail with a group of young people who have earned a grant from the Ministry to take this supposedly educational trip.

During the first night of the trip a world catastrophe causes a cataclysm that makes the earth disappear. After the disaster, the ship's crew are isolated in the high seas and commence a series of adventures aimed at finding the little earth that, they think, might still exist.

Analysis Methodology

For the analysis of the series the qualitative methodological framework developed by the group UNICA from the Pompeu Fabra University (Ruiz-Collantes, Ferrés, Obradors, Pujadas and Pérez 2006) was used. The system is based on Greimas' semiotic framework (1989) that enables the most important features of the audiovisual product to be identified from the object of value, which is the starting point of the storylines (Lacalle 2013; Luzón et al. 2011; Vives 2007).

The analysis framework is centred on the narrative roles that a character may perform as seen on *Figure 1*.

When the character is situated within a narrative role other aspects of this character may be identified, such as physical and emotional features, what makes him/her act, his/her beliefs, objectives, contact with and attitude towards other characters, the results of his/her objectives, etc.

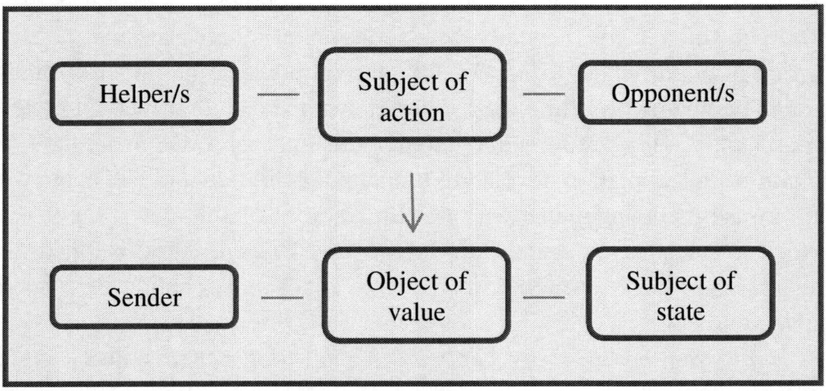

Source: own creation from Ruiz-Collantes et al. (2006).

Figure 1. Narrative programme – narrative roles.

In order to be able to analyse the relationships represented in the fictional series it is important to find out what drives the characters to act, to carry out their interventions, the skills the characters apply to develop actions, the type of actions they perform, etc. The construction of the image of the love relationship is closely related to the actions and motivations of the characters. Thus, the image of the love relationship constructed within the series is not the same if, for example, the character performing the action in the romantic storyline is a boy or a girl, whether the action ends successfully or not, etc. Also important is the narrative role that boys and girls fulfil within the love relationship compared to the rest of the narrative roles that appear in the storylines, that is, when for example the girl adopts the role of subject of action in the storyline, what objectives is she seeking? Does anyone help her achieve them? Does anyone oppose her objectives? When, on the other hand, it is the boy who is adopting the role of subject, what situation occurs? Are his objectives different from those of the girl? Does his action end successfully? This information is important for completing the image constructed upon the love relationship as it enables it to be conceived in a broader area of action. This framework therefore enables us to obtain information about the characters and, from them, about the representation of the love relationships that comprise the series.

For the purposes of our study, an analysis was made of the storylines about love relationships in which the most popular characters were involved.

In the case of *Los Protegidos,* the relationship between Culebra and Sandra (5 storylines analysed); and for *El Barco*, the relationship between Ainhoa and Ulises (5 storylines analysed). These data were extracted from a prior survey that was given out to adolescents of different ages (Masanet 2015).

Semiotic Analysis of Los Protegidos and El Barco

Los Protegidos

Los Protegidos presents us with the impossible love between the two young main characters, Sandra and Culebra, who are attracted to each other when they first meet but who cannot touch each other because of Sandra's special electrical powers. Sandra gradually manages to control this power and is able to start touching other members of the family and her friends, but this does not happen with Culebra. As she is very much in love she cannot control her special powers when she is near him.

The semiotic analysis reveals that *Los Protegidos*, through the relationship between Culebra and Sandra, is presenting us with a relationship based on romantic love (Sternberg 1989). We encounter a relationship that does not achieve long-term commitment due to an external factor that prevents it: Sandra's special powers. This is why they are forced to love each other without being able to touch one another. At certain points in the series this results in Sandra and Culebra trying out relationships with third parties that always end up failing as they are not with their "true love." Taking as a reference the study by Balló and Pérez (1995) on cinema's universal plots, we might categorize Sandra and Culebra's relationship as a *forbidden love* relationship. Culebra and Sandra represent the modern day Romeo and Juliet who live in the society of individualism and, due to external factors, cannot be together. In this case the external factors are not hatred between families, but rather Sandra's special powers that prevent physical contact between the lovers. The impossible love storyline exists in the relationship between Sandra and Culebra with a modern twist. Also important is the fact that there is no tragic finale; instead they manage to overcome the prohibition and consummate their love. As this is a series for all members of the family, it is logical that the relationship between Culebra and Sandra is finally consummated after three seasons of waiting, in order to reward the audience with a happy ending.

Throughout the series we see how the relationship between Sandra and Culebra perpetuates many of the romantic love myths: a) "the power of love"; b) "predestined true love" and c) "the total gift of love" (de la Peña, Ramos, Luzón and Recio 2011; Ferrer, Bosch, Ramis, Torres and Navarro 2006; Luzon, Ramos, Recio and de la Peña 2011). As mentioned above, Sandra and Culebra's relationship is presented as a forbidden love that in the end reaffirms the idea that "love conquers all" –the power of love– as, in the end, the characters manage to be together. This myth is associated with the idea that the force of love is enough to change the other person (a myth that is proven to be highly dangerous such as in situations of domestic violence). Love is conceived as something superior that enables everything to be overcome; it ratifies "*the power of love*." Likewise, the compatibility between love and suffering is also affirmed – there's never one without the other - , which leads us to the standardization of the idea that 'loving is suffering.' These characteristics reaffirm the inexorable statement that true love can conquer all, as it changes people, overcomes conflicts, etc. Thus, true love is a love that is able to forgive everything. The relationship between Sandra and Culebra consistently invokes this myth and reaffirms it with its happy ending, as the characters achieve their final object of value, being able to be together and being able to touch each other.

The myth of "*predestined true love*" is also exploited. The idea that there is one person for each of us, a person that is unique, is reaffirmed. Sandra and Culebra cannot forget each other because it is their destiny to be together. This is why attempts at relationships with other characters in the series always fail in the end. They are unsuccessful because they are not with the predestined person. The myth of "predestined true love" refers to the platonic ideal of one's other half and, therefore, our complementarity. Without the other half we are not a unit, we are incomplete in life. This idea suggests that there is only one "true love" in life, and that if we let it get away we will never find it again. We cling to what we believe is our soul mate and, as it is true love, this feeds the illusion of eternal passion.

The relationship between Sandra and Culebra also manifests the myth of "*the total gift of love*." The main characters cannot live without each other so will do everything they can to overcome the problem of their special powers and thus consummate their love. This myth refers to the idea that love is the essence of life: the fundamental element. This means you cannot live without the person you love, the one who is your true love because happiness is linked to this person. Without love there is no happiness. Given that you cannot feel happiness without the love from the other person, the need is invoked for this total gift in order not to lose it. You do everything for love, even change the way you are and renounce personal intimacy.

Along with this assertion of romantic myths, we also encounter stereotypical operations in terms of gender. Culebra has relationships with other characters in the series, relationships that we are given to understand include sex. Meanwhile, Sandra never has a relationship with another person in the series; she only manages to get close, in friendship terms, to other male characters. This representation appears to relate to a gender stereotype whereby boys are allowed to have more relationships than girls. Sandra must remain pure for Culebra, while he can diversify with other girls.

Thus, we see that *Los Protegidos* presents an adolescent relationship based on the concept of romantic love. Through forbidden love the series invokes and reaffirms different myths within this forbidden love.

El Barco

In the case of *El Barco* we analyse a relationship that in appearance is very different from that of *Los Protegidos* but, as we will see later on, shares common points. Ainhoa and Ulises meet each other on the ship. She has been given a grant to take the trip while he is a stowaway who is hiding in order to try and meet his real father, one of the crew members. They feel attracted to each other right from the start but they start relationships with other people on board. Over time they realise they want to be together but their relationship will not be easy and they come up against various obstacles that hinder it.

The semiotic analysis shows that the series invokes the "myth of true love as something that is predestined." Ainhoa and Ulises are predestined to be together and, therefore, the other relationships they instigate will be condemned to fail. Despite the fact that their relationships with other characters start passionately, this soon disappears and they become monotonous and void of any emotional intimacy. Subsequently, while they maintain their respective partners, a passion arises between Ainhoa and Ulises. They do not know each other, so they do not have any kind of intimacy, but they do feel an intense passion and attraction for each other. Despite trying to avoid it in order not to breach their commitment to their respective partners, the passion they feel for each other is too strong and they end up finishing their other relationships in order to be together. Thus the series shows us the irrationality of passion as a fundamental ingredient in love relationships.

When the characters finally leave their respective partners and start a relationship they begin to become emotionally intimate. This situation does not last long however as Ainhoa's former partner –Gamboa- threatens her if she stays with Ulises and she is forced to break off her relationship with him. At this point, as in *Los Protegidos*, we encounter a "*forbidden love*" relationship (Balló and Pérez 1995). Again, the story of Romeo and Juliet is represented in the modern day; a love that cannot be consummated due to an external factor (Balló and Pérez 1995), in this case, the threats by Gamboa. The story reaches a turning point when Ainhoa breaks the prohibition, stands up to Gamboa's threats and consummates her relationship with Ulises. Yet the situation is short lived as Gamboa uses his ammunition to make Ainhoa see that she does not really know Ulises, that she is in an irrational love (Sternberg 1989) where there is passion and commitment but no emotional closeness. Ainhoa does not really know Ulises because they have based their relationship on the passion they feel for each other. Therefore, when Gamboa provokes Ulises to violence in front of Ainhoa, she breaks off the relationship with him. Ainhoa feels disappointed in Ulises' behaviour and concedes that she does not really know him.

Yet the fervour they feel continues to be so strong that it brings them together and, once again, they start a "passionate" relationship based on what they feel but without taking into account the need to construct a space for emotional intimacy.

This relationship is maintained until Ainhoa discovers new secrets about Ulises and once again loses her trust in him. Ainhoa is again disappointed by not really knowing the person she is having a relationship with and they split. Despite the fact that they do not see each other for a long time and, therefore, they do not have the chance to acquire the intimacy that is lacking in their relationship and that it needs, at the end of the series we see that she decides that she is in love with Ulises and that she wants to be with him. Ainhoa, individually, without having seen Ulises again since they put their relationship on hold, acquires the commitment to go back to him as irrational passion has won her over.

Nevertheless, the series holds a surprise. We do not know whether Ainhoa and Ulises end up consummating their relationship as Gamboa shoots Ulises and the series does not reveal whether the character lives or dies. In the final image we see Ainhoa dressed as a bride but we do not know whether she is marrying Ulises or not.

Synthetically, we might state that the character of Ainhoa is in a risk situation caused by the blind trust in her irrational passion that does not enable her to evaluate and reflect on Ulises' true character. At the same time, this risk situation is framed within a relationship of "romantic love" that incorporates different myths such as that of "predestined true love," "the power of love" and "the other half" -myths that are predominantly reflected in the characters' storyline throughout practically the whole series; a story of "forbidden love"-. As in *Los Protegidos*, in the modern day it would be unusual for the relationship to be prohibited by the family, so it is the result of external factors. This factor is initially the threat of Gamboa. When Ainhoa breaks the prohibition –his threats–, she overcomes the external factor and goes back to Ulises. The series then needs other factors to keep them separate, to prevent their relationship, to be able to keep maintaining the myth of "forbidden love." At this point it starts to use the resource of lack of intimacy to keep the couple separate. The series' storyline constantly reveals that the characters do not know each other, that there are secrets between them and this is what stops them from being together. It is always the female character, Ainhoa, who offers up full intimacy but she does not receive it from the male character. Ulises keeps secrets that cause breakups on different occasions and make their relationship impossible. Despite the lack of intimacy, there is an attempt to unite the couple as at the end of the series Ainhoa decides that she wants to be with Ulises, even though they have not seen each other for months and they have not been able to build up intimacy.

Given the nature of the relationship represented in the series, throughout the analysis we saw how the object of value in the storylines of the love relationship between Ulises and Ainhoa is consistently to conquer the other, to overcome the obstacles that Gamboa puts in their way, to achieve commitment, etc. Overall, it can be stated that the objective of the two characters always revolves around consolidating the relationship, which is impossible, as they represent a relationship of "forbidden love" that would not make sense without the factor of the impossibility. Lastly, it is important to note that this relationship is associated with many of the myths of "romantic love," in the same way as in *Los Protegidos*.

CONCLUSION

Jónasdóttir (1994) wondered why in countries with significant development in equality policies, such as the Nordic countries, there are still many differences in the relationships between men and women. One of the conclusions reached is that classical forms of patriarchy have evolved into more modern forms where love ends up being a manifestation of power relations yet still with gender differences. In Spain, the woman remains in charge of the great emotional sphere, family and home and the man holds the weight of social representativeness. The man also takes advantage of the work charged to women to care about and tend to love relationships.

As different authors have demonstrated in the twentieth century (Bauman 2005; Beck and Beck-Gernsheim 2001; Giddens 2008) a link between "romantic love," "love passion" and "sexuality" has been constructed. Nowadays -in the West- the love relationship is related to a free and individual choice based on an emotional state of infatuation. Looking at the great revolution that has taken place in matters of love relationships, it seems that this freedom of choice grants equality to the two members in the relationship, or at least it should. Giddens (2008) reflects this when he says that true intimacy is the consequence of the democratization process in healthy interpersonal relationships. Yet other studies show how society, especially among girls, has internalized an important set of stereotypes and "false beliefs" about what "true love" is and the "feminine" capacity to love. Furthermore, the media play an important role in this context. The media present certain messages about the meaning of love relationships and gender roles within relations. In a television series when a mother tells her daughter that "love is suffering," certain myths and stereotypes that may become dangerous are being represented.

Our research is situated in this context. The study is based on the relationship between media culture and adolescent culture. The use of the media is part of young people's everyday activities including personal relationships, domestic environment, interactions with peers and various other contexts. Therefore, an analysis of the representations of love relationships in the media is important. As we stated in our analysis, the series wrote about present loving relationships based on certain stereotypes that perpetuate gender differences. These series continually appeal to different myths of romantic love such as "the power of love," "the predestined true love" and "the love as possession."

This trend is in line with the results of previous research (Aran et al. 2011; Capdevila et al. 2011; Luzon et al. 2011). These myths are easily recognized by viewers as they have a significant presence in contemporary society (Ackerman 2000; Esteban 2011; Medina et al. 2007) and may be very impoverishing.

Our final conclusion is that old socio-cultural models are still prevalent in new TV formats. An idea that stands out from our analysis is that the set of romantic myths is in fact furnishing a meta-myth of a higher order that views romantic love as forbidden love. What remains of the new Romeos and Juliets is the continued existence of love relationships based on emotional inequality between men and women, where the idealization of love as a forbidden dream covers up the risks of abusive relationships. Thus, the new meta-myth of forbidden love encompasses all the previously noted romantic myths - "the power of love," "predestined true love" and "the total gift of love."

ACKNOWLEDGMENT

This chapter is based on a doctoral dissertation entitled "Media representation and adolescent interpretation of sexuality and the love relationship in fictional series." This dissertation is part of the RandDandI project funded by the Ministry for Economy and Competitiveness of Spain – key code EDU2010-21395-C03– entitled "Undergraduate teaching faced with the audiovisual communication competence in a digital environment."

REFERENCES

Ackerman, D. (2000). Una historia natural del amor [A natural history of love]. Barcelona, Spain: Anagrama.

Alexander, A. (1985). Adolescents' soap opera viewing and relational perceptions. *Journal of Broadcasting and Electronic Media,* 29(3), 295–308.

Aran, S., Medina, P., Munté, R.-A., and Rodrigo, M. (2011). Joves, amor i sèries de televisió. Incidència de l'alfabetització audiovisual en la (re)interpretació dels relats amorosos televisius [Young people, love and television series. Incidence of media literacy in the (re)interpretation of TV love stories]. *Quaderns Del CAC*, 36, 115–123.

Babrow, A. S. (1987). Student motives for watching soap operas. *Journal of Broadcasting and Electronic Media, 31*(3), 309–321.

Badmington, N. (2004). Roswell High, Alien Chic and the In-Human. In G. Davis and K. Dickinson (Eds.), Teen TV: genre, consumption and identity (pp. 166–175). London, UK: British Film Institute.

Balló, J., and Pérez, X. (1995). *La semilla inmortal. Los argumentos universales en el cine* [The immortal seed. Universal arguments in the cinema]. Barcelona, Spain: Editorial Anagrama, S.A.

Banks, M. J. (2004). A Boy for All Planets: Roswell, Smallville and the Teen Male Melodrama. In G. Davis and K. Dickinson (Eds.), Teen TV: genre, consumption and identity (pp. 17–28). London, UK: British Film Institute.

Barker, C., and Andre, J. (1996). Did you see? Soaps, teenage talks and gendered identity. *Young: Nordic Journal of Youth Research, 4*(4), 21–38.

Bauman, Z. (2005). Amor líquido. Acerca de la fragilidad de los vínculos humanos [Liquid love: on the frailty of human bonds]. Buenos Aires, Argentina: Fondo de Cultura Económica.

Bay-Cheng, L. Y. (2003). The Trouble of Teen Sex: The construction of adolescent sexuality through school-based sexuality education. *Sex Education: Sexuality, Society and Learning,* 3(1), 61–74.

Beck, U., and Beck-Gernsheim, E. (2001). El normal caos del amor: las nuevas formas de la relación amorosa [The normal chaos of love]. Barcelona, Spain: Editorial Paidós.

Bragg, S. (2006). "Having a real debate": using media as a resource in sex education. *Sex Education: Sexuality, Society and Learning,* 6(4), 317–331.

Buckingham, D., and Bragg, S. (2004). Young people, sex and the media. Basingstoke, UK: Palgrave.

Capdevila, A., Crescenzi, L., and Araüna, N. (2013). Relaciones afectivas, adolescencia y series de ficción. Sexo y amor en Sin tetas no hay paraíso [Relationships, adolescence and fiction series. Sex and love in Sin tetas no hay paraíso]. *Miguel Hernández Communication Journal,* 9(47), 191–212.

Capdevila, A.; Tortajada, I., and Araüna, N. (2011). Els rols de gènere, les relacions d'amor i de sexe en les sèries de ficció. El cas de Sin tetas no hay paraíso [Gender roles and relationships of love and sex in fiction series. The case of Sin tetas no hay paraíso]. *Quaderns del CAC,* 13(36), 63-69.

Coltrane, S., and Messineo, M. (2000). The perpetuation of subtle prejudice: race and gender imagery in 1990s television advertising. *Sex Roles,* 42, 363–389.

Crespo, M. (2005). Mensajes y modelos televisivos para los adolescentes: estudio base para un análisis sistemático del contenido sexual de las series de televisión [Messages and television models for teens: study for a systematic analysis of sexual content of television series]. *Doxa,* 3, 187–214.

Davis, D. M. (1990). Portrayals of women in prime-time network television. *Sex Roles*, 23, 325–322.

Davis, G. (2004). “Saying It Out Loud”: Revealing Television’s Queer Teens. In G. Davis and K. Dickinson (Eds.), Teen TV: genre, consumption and identity (pp. 127–140). London, UK: British Film Institute.

de la Peña, E. M., Ramos, E., Luzón, J. M., and Recio, P. (2011). Andalucía Detecta. Sexismo y violencia de género en la juventud [Sexism and gender violence in youth]. Sevilla, Spain: Instituto Andaluz de la Mujer.

Douglas, K., and William, K. M. (2004). “We Don”t Need No Education’: Adolescents and the School in Contemporary Australian Teen TV. In G. Davis and K. Dickinson (Eds.), Teen TV: genre, consumption and identity (pp. 151–165). London, UK: British Film Institute.

Erikson, E. (1980). Identidad: juventud y crisis [Identity: youth and crisis]. Madrid, Spain: Taurus.

Erikson, E. (1972). Sociedad y adolescencia [Society and adolescence]. Buenos Aires, Argentina: Editorial Paidós.

Esteban, M. L. (2011). Crítica del pensamiento amoroso [Criticism of loving thought]. Barcelona, Spain: Ediciones Bellaterra.

Fedele, M. (2011). El consum adolescent de la ficció seriada televisiva [The adolescent consumption of television fiction series]. (Doctoral Thesis, Universitat Autònoma de Barcelona, Spain). Retrieved from http://www.tdx.cat/handle/10803/83502.

Fejes, F. J. (1992). Masculinity as fact. A review of empirical mass communication research on masculinity. In S. Craig (Ed.), Men, masculinity, and the media, (pp. 9–22). London, UK: Sage.

Ferrer, V. A., Bosch, E., Ramis, C., Torres, G. and Navarro, C. (2006). La violencia contra las mujeres en la pareja: creencias y actitudes en estudiantes universitarios/as [Domestic violence: Beliefs and attitudes in university students]. *Psicothema,* 18(3), 359-366.

Fernández-Cavia, J. (2002*). El consumidor adolescent: Televisió, marques i publicitat* [The teenager consumer: Television, brands and advertising]. Barcelona, Spain: Aldea Global.

Galician, M. L. (2004). Sex, love, and romance in the mass media: Analysis and criticism of unrealistic portrayals and their influence. Philadelphia, USA: Lawrence Erlbaum Associates.

García-Muñoz, N., and Fedele, M. (2011). Las series televisivas juveniles: tramas y conflictos en una “teen series” [Television fiction series targeted at young audience: plots and conflicts portrayed in a teen series]. *Comunicar,* 19(37), 133–140.

Giddens, A. (2008). La transformación de la intimidad: Sexualidad, amor y erotismo en las sociedades modernas [The transformation of intimacy: Sexuality, love and eroticism in modern societies]. Madrid, Spain: Cátedra.

Greimas, A. J. (1989). Del sentido II. Ensayos semióticos [Sense II. Semiotic trials]. Madrid, Spain: Editorial Gredos.

Guarinos, V. (2009). Fenómenos televisivos "teenagers": prototipias adolescentes en series vistas en España [Televisual teenager phenomena. Adolescent prototypes in TV series in Spain]. *Comunicar,* 17(33), 203–211.

Hetsroni, A. (2008). Dependency and Adolescents' Perceived Usefulness of Information on Sexuality: A Cross-Cultural Comparison of Interpersonal Sources, Professional Sources and the Mass Media. *Communication Reports,* 21(1), 14-32.

Hills, M. (2004). Dawson's Creek: "Quality Teen TV" and "Mainstream Cult"? In G. Davis and K. Dickinson (Eds.), Teen TV: genre, consumption and identity (pp. 54–67). London, UK: British Film Institute.

Huertas A., and França E. (2001). El espectador adolescente. Una aproximación a cómo contribuye la televisión en la construcción del yo [The teenager viewer. An approach to how tv helps in building the identity]. *ZER: Revista de estudios de comunicación,* 11, 331–350.

Jónasdóttir, A. (1994). Why women are oppressed. Philadelphia, USA: Temple University.

Kirby, D. (2002). The impact of schools and school programs upon adolescent sexual behavior. *Journal of Sex Research,* 39(1), 27–33.

Lacalle, C. (2013). Jóvenes y ficción televisiva. Construcción de la identidad y transmedialidad [Young people and television fiction. Construction of identity and transmediality]. Barcelona, Spain: Editorial UOC.

Luzon, J. M., Ramos, E., Recio, P., and de la Peña, E. M. (2011). Proyecto Detecta Castilla y León. Factores de riesgo y de protección en la prevención contra la violencia de género en la pareja. Un estudio de investigación en la población adolescente de Castilla y León [Risk and protection factors in the prevention against gender violence in the couple. A research study in the adolescent population of Castilla y León]. Junta de Castilla y León, Spain: Consejería de familia e igualdad de oportunidades.

Luzón, V., Ramajo, N., Figueras, M., Capdevila, A., Gómez, L., Jiménez, M., and Ferrer, I. (2011). La Imagen de los/as Adolescentes en el Prime Time Televisivo. Transmisión, consumo y recepción [Image of the adolescents in prime time television. Transmission, consumption and reception]. Spain: Ministerio de trabajo y asuntos sociales. Instituto de la Mujer.

Masanet, M.-J. (2015). Representació mediàtica i interpretació adolescent de la sexualitat i la relació amorosa en la ficció seriada [Media representation and adolescent interpretation of sexuality and the love relationship in fictional series]. (Doctoral Thesis, Universitat Pompeu Fabra, Spain). Retrieved from http://www.tdx.cat/handle/10803/292732.

Masanet, M.-J., and Buckingham, D. (2015). Advice on life? Online fan forums as a space for peer-to-peer sex and relationships education. *Sex Education: Sexuality, Society and Learning,* 15(5 - Special Issue: Entertainment Media's Evolving Role in Sex Education), 486-499.

Masanet, M.-J., Medina, P., and Ferrés, J. (2012). Representación mediática de la sexualidad en la ficción seriada dirigida a los jóvenes. Estudio de caso de Los Protegidos y Física o Química [Media representation of sexuality in fiction series aimed at young people. Case study of Los Protegidos and Física o Química]. *Revista Comunicación,* 10(1), 1537–1548.

Medina, P., Aran, S., Munté, R.-A., Rodrigo, M., and Tharrats, J. (2007). Violència simbòlica i models amorosos en la ficció televisiva seriada per al consum adolescent i juvenil. Estudi de cas (Porca misèria) [Symbolic violence and love models in tv fiction for adolescent and young consumption. Case study (Porta misèria)]. Barcelona, Spain: Quaderns Del CAC.

Medina, P., and Rodrigo, M. (2009). Análisis de la estructura narrativa del discurso amoroso en la ficción audiovisual. Estudio de caso: "Los Serrano" y "Porca Misèria" [Analysis of the narrative structure of amorous discourse in audiovisual fiction. Case Study: "Los Serrano" and "Porca Misèria"]. *ZER: Revista de Estudios de Comunicación,* 14(27), 83–101.

Moran, K. C. (2003). A reception analysis: Latina teenagers talk about telenovelas. *Global Media Journal-American Edition,* 2(2).

Pindado, J. (2005). Resultados de un estudio con jóvenes de Málaga. Los medios de comunicación en la sociedad adolescente [Results of a study with young from Málaga. The media in the adolescent society]. *Telos,* 62, 14–20.

Ruiz-Collantes, X., Ferrés, J., Obradors, M., Pujadas, E. and Pérez, O. (2006). La imatge pública de la immigració a les sèries de televisió [The Public Image of Immigration in TV series]. *Quaderns del CAC,* 23-24, 103–126.

Simelio, N. (2010). La representación de las relaciones sociales en las series de ficción digitales creadas específicamente para Internet. La televisión como contribución a la alfabetización digital [The representation of social relationships in digital fiction series created specifically for the Internet. Television as a contribution to digital literacy]. In I Congreso Euro-Iberoamericano Alfabetización Mediática y Culturas Digitales [I Euro-Iberoamerican Congress on Media Literacy and Digital Cultures]. Sevilla, Spain.

Sternberg, R. J. (1989). El triángulo del amor. Intimidad, pasión y compromiso [The triangle of love. Intimacy, passion and commitment]. Barcelona, Spain: Ediciones Paidós Ibérica.

Vives, M. (2007). La representació dels adolescents en les telenovel•les de producció pròpia de TV3 [The representation of teenagers in the self-produced telenovelas of TV3]. (Master's thesis, Universitat Pompeu Faba, Spain).

Von-Feilitzen, C. (2004). Young people, soap operas and reality TV. Götenborg, Sweden: Götenborg University Press.

Ward, L. M. (1995). Talking about sex: Common themes about sexuality in the prime-time television programs children and adolescents view most. *Journal of Youth and Adolescence,* 24(5), 595–615.

In: Television and Romance
Editor: Amir Hetsroni

ISBN: 978-1-63485-077-3

Chapter 3

IS THERE A 'HAPPILY-EVER-AFTER' FOR THE CHILDFREE? AN ANALYSIS OF CHILDFREE WOMEN ON TELEVISION

Andrea McClanahan
East Stroudsburg University of Pennsylvania, PA, US

ABSTRACT

This chapter attempts to interrogate fictional representations of childfree women on television and how the hegemonic ideal of motherhood is framed in relation to these women. An analysis of the following characters is conducted to determine the frames in which childfree women are placed in fictional television series: Whitney from *Whitney* (2011-2013), Bernadette from *Big Bang Theory* (2007- present), Carrie from *Sex and the City* (1998- 2004), Robin from *How I Met Your Mother* (2005-2014), and Cristina from *Grey's Anatomy* (2005- present). The main argument is that there are two specific frames used for childfree women.

The first is "romance trumps all" where women who verbally state their desires to not have children are easily swayed when presented with a choice to end a romantic relationship or change their opinion on their stated aspirations in relation to children. The second, "childfree by choice, romance-free by consequence," presents women who are committed to being childfree regardless of the consequence. In this frame, the women are often left without romantic relationships due to their childfree status.

INTRODUCTION

In the introduction to her book, *Conceiving normalcy: Rhetoric, law, and the double-binds of infertility*, Elizabeth Britt (2001) writes, "Why do people pursue parenthood? To many the answer is so obvious that the question seems unnecessary: to want a child is a natural, normal desire" (p. 1). The "norm" in our society is for individuals to grow up, fall in love, get married, and have children. This is the life script told to us through media images and cultural narratives.

However, more women and men are choosing not to have children and remain childfree[1]. United States Census Data from 2010 shows that the percentage of childfree women between 40-44 years of age is around 19 percent (Tavernise 2011).

This number shows an 80% increase since 1976 when the percentage of women who had never given birth between the ages of 40-44 was 10 percent (Livingston 2010). Livingston further explains that the number of childfree women has increased in recent decades because the "social pressure to bear children appears to have diminished for women and that today the decision to have a child is seen as an individual choice. Improved job opportunities and contraceptive methods help create alternatives for women who choose not to have children.

From the *Time* magazine cover story, "The childfree life: When having it all means not having children" (Sandler 2013) to Melanie Notkin's (2014) book, *Otherhood: Modern women finding a new kind of happiness*, or Henriette Mantel's (2013) edited collection, *No kidding: Women writers on bypassing parenthood*, the childfree revolution is finally receiving some much deserved press.

However, despite the increase in women not having children and the publication of articles, books, and even blogs like *TheNotMom.com* popping up and gaining popularity, the decision to be childfree—whether by choice or circumstance—has largely gone unrecognized in fictional television programming.

In this chapter, I argue that television depictions of childfree women are difficult simply because they do not allow for the following of a traditional

[1] Throughout this chapter, the term "childfree" is used when referring to women who do not have children. The preference for the term childfree over childless is one that has been debated among various scholars. The term "childless" implies that women who do not have children are less-than their childbearing counterparts. Further, "childless" emphasizes the idea that women naturally want children which is not always the case. (Cain 2001).

storyline—one that involves love and romance, marriage, and children. Due to the challenging nature of this storyline, there are very few childfree characters on television and only a small percentage of those have openly discussed their desire for remaining childfree or their reasons for not having children. May (1995) posits that "People reconcile their desires, felt needs, inclinations, and biology with social norms and pressures in a wide variety of ways. In a society that expects most adults to have children, their feelings about their status as parents or nonparents are uniquely their own" (p. 2).

While the personal nature of the decision not to have children may be a contributing factor to the lack of storylines devoted to childfree female characters, there is an augmented need for representations of childfree women (and men) on television as the number of people remaining childfree increases in our society. Kellner (2015) argues:

> Radio, television, film, popular music, the Internet and social networking, and other forms and products of media culture provide materials out of which we forge our very identities, including our sense of selfhood; our notion of what it means to be male or female; our conception of class, ethnicity, and race, nationality, sexuality; and division of the world into categories of 'us' and 'them.' Media images help shape our view of the world and our deepest values; what we consider good or bad, positive or negative, moral or evil. Media stories provide the symbols, myths, and resources through which we constitute a common culture and through the appropriation of which we insert ourselves into this culture. Media spectacles demonstrate who has power and who is powerless, who is allowed to exercise force and violence and who is not. They dramatize and legitimate the power of the forces that be and show the powerless that they must stay in their places and be oppressed (p. 7).

With the lack of media representations of childfree women, the decision to be childfree remains one outside of the cultural norm for what is expected in our lives—a hegemonic ideal.

This calls for an explanation of the term hegemony as done by Corcoran (1984). "Hegemony implies the active engagement of individuals with the ideology of the dominant sectors of society and therefore active cooperation in their own domination" (p. 142). Women who automatically assume they must have children in order to fit what is expected of them are colluding with the hegemonic ideal of romance, love, marriage, and children because as Lull (2015) argues "information and entertainment technology is so thoroughly integrated into the everyday realities of modern societies, mass media's social influence is not always recognized, discussed, or criticized. Hegemony, therefore, can easily go undetected" (p. 40). In order for us understand the possibilities outside of the hegemonic ideal, we must realize that to a certain extent "women's desires are the result of bodily inscription and of metaphors and story lines that catch them up in ways of being/desiring from which they have no escape unless they can reinscribe, discover new story lines, invert, invent, and break the bounds of the old structures and old discourses" (Davies 1992, p.58).

To draw attention to the hegemonic narrative implying that women must have children in order to have a successful life, this chapter focuses on contemporary female television characters[2] who make explicit statements at some point in the series indicating their lack of desire to have children and how this decision impacts their romantic relationships[3]. Specifically, I analyze how childfree women are framed in television narratives by asking two questions:

How do television writers frame a female character who desires a life without children?

What is the impact of this decision on the character's romantic relationships?

I examine the framing of characters, which – for the purpose of this chapter – can be understood as a way in which particular meanings are imposed upon situations, events, or issues; frames can be manipulated to convey particular desired meanings by actors and institutions (Jaworski 2009).

After analyzing five female television characters—Whitney from *Whitney*, Bernadette from *Big Bang Theory*, Carrie from *Sex and the City*, Robin from *How I Met Your Mother*, and Cristina from *Grey's Anatomy*, it is clear that characters who explicitly express their desires not to have children are framed in two ways. The first frame is the perspective that when women are confronted with a decision about having children that will impact their romantic relationships, romance trumps all. The second frame includes women who adhere to their decision to remain childfree and, as a consequence,

[2] Additional representations of childfree female television characters can be found in shows such as Cheers, The Bob Newhart Show, Laverne and Shirley, The Andy Griffith Show, Seinfeld, Sex and the City (Samantha) and The Mary Tyler Moore Show.

[3] While characters such as Elaine Bene in Seinfeld and Samantha Jones in Sex and the City comment on their opinions about having children, with Samantha even going as far as to throw an "I'm not pregnant party" (Minsky and Seidelman, 1998) their decision is never a plot point for romantic relationships and will not be covered in this chapter.

struggle with romantic relationships—often leaving them romance-free as well as childfree.

Romance Trumps All

One of the most common ways in which television deals with childfree women is to put them in situations where they need to make a decision about whether they want to have children in order to sustain a romantic relationship. This frame presents women as having the ultimate choice to make—one that determines the state of a romantic relationship based upon whether the woman chooses to follow the hegemonic ideal of motherhood.

The television sitcom, *Whitney,* created by comedienne Whitney Cummings and aired for two seasons (2011-2013), had focused on a cohabitating couple that has been living together for three years (Whitney 2015). Throughout the beginning of the first season of the show, it is clear that Whitney does not want to get married or have children. In an early episode, Whitney asks Alex, her boyfriend, to marry her. He responds by recognizing that she does not really want to get married so he says no (Cummings and Ackerman 2011).

Whitney acknowledges that her romantic relationship with Alex is just as committed as individuals who are married and even makes the argument in the episode "Clarence!" when she and Alex attempt to adopt a puppy. When she realizes that the form she is filling out to adopt a puppy only has the options of "single" or "married," she approaches the desk and exclaims to Mrs. Janks, the co-owner of the shelter, "Your form is broken. You don't have a box for what our relationship is." The woman at the desk responds, "Oh, he won't propose?" Whitney, clearly irritated, says "Oh, we aren't married but we've been together for three years." She continues, "If I had a soul, he would be its' mate. During Mrs. Janks' home visit, she criticizes Whitney and Alex saying, "I hope you're not planning on having any kids because you can't handle a new baby and a new dog at the same time." Alex responds, "Oh no, trust me. I smoked so much weed so it's like I have no sperm" (Sklaver and Cendrowski, 2011). With that interaction, the audience is left to think that the discussion of potential children in the relationship between Whitney and Alex is comic and one that is not probable.

While Whitney and Alex have a romantic relationship and many episodes are devoted to the two attempting to create over-the-top romantic situations, children do not appear to be part of their romantic relationship equation.

However, in the episode, "Up All Night" Whitney's desire not to have children is dealt explicitly. When her friends are in her apartment and Roxanne, one of her close friends, mentions having her eggs frozen so she can have children in the future, Whitney exclaims, "Okay. Nope. No. We do not do this. We do not sit around and talk about having kids, okay? Come on. We're sluts" (Cummings, Quaintance and Ackerman 2011). Later, when Whitney buys a pair of baby shoes while under the influence of a sleeping pill, she and Alex go to her therapist to discuss what it means.

WHITNEY: Kids are not part of our thing. I'm not thinking about kids. He's not thinking about kids.

ALEX: How do you know? We never even talked about it.

WHITNEY: That's because neither of us wants kids.

ALEX: No, I want kids. We haven't talked about it because I was afraid that you don't and [hesitates]

WHITNEY: What?

ALEX: It's kind of a deal-breaker.

The audience is left thinking that Whitney does not want children after this brief conversation because she states it explicitly to her therapist and Alex. However, the conversation continues when the two return home.

ALEX: This is a really bad conversation to have. Because if you never want kids [pauses], it's just a really bad conversation to have. Okay?

WHITNEY: Okay. I didn't say I never want kids. I just, I grew up around a lot of divorce and it sucks and I would never want to do that so I would have to make sure that would never happen to us.

ALEX: Well, in order to get divorced, you have to get married.

WHITNEY: Duh.

ALEX: Well, does that mean you would consider getting married?

WHITNEY: To you? I probably would.

ALEX: Oh my god! Whit, you're like a person!

WHITNEY: Shut up. No, I'm not. I'm scared of marriage and I'm scared of having kids but with you it's like the first time I can actually imagine it. (Cummings, Quaintance and Ackerman 2011)

The implication of Alex's statement "You're like a person" is extremely telling in our culture when it comes to what is expected for a woman to be considered normal in our culture. Women must get married and have children to be real people. Grose (2011) writes, "Whitney is meant to be an independent woman, one who gets joy out of standing up for herself, giving her boyfriend the business, and making jokes about balls. So what does she do when she learns that kids are a deal-breaker for her boyfriend? She completely folds. What's more, Whitney's concession toward marriage and babies is seen as a fully positive development in the show."

Whitney had a choice to make—consider children as a part of her future or end her relationship with her long-time boyfriend. Whitney chose what society expects women to choose—have children in order to sustain a romantic relationship.

Another character that has shown explicit disdain for having children is Bernadette from *Big Bang Theory.* The *Big Bang Theory* is a television show that revolves around a group of nerds and their friendships and romantic relationships. Bernadette is in a relationship—first a romantic relationship and then marriage—with Howard, one of the main characters on the show.

Early in their relationship, Howard discovers that Bernadette does not like children when he attempts to have her help him with a magic show at a children's birthday party. Bernadette explains to Howard that she may not be the best choice for an assistant because "I'm not that comfortable with little kids" (Lorre, Holland, Hernandez and Cendrowki 2012). Howard quickly dismisses her concerns. However, later in the episode, Howard experiences Bernadette's interactions with children as she reprimands them for interrupting Howard's magic show saying "That's it. No cake for you. Anyone else want to join the No Cake Club?" (Lorre, Holland, Hernandez and Cendrowki 2012).

At the end of the party, Bernadette explains to Howard that she does not like children because she served as the primary caretaker for her brothers and sisters while her mother was at work.

BERNADETTE: I'm sorry. I know it makes me sound like a bad person, but I just don't like children.

HOWARD: Yeah, no, we all got that. But don't you think it'll be different when the child is ours?

BERNADETTE: Right, when it's our kid that's ruined my body and kept meup all night and I've got no career and no future and nothing to be happy about for the next 20 years? Sure, that'll be completely different.
(Lorre, Holland, Hernandez and Cendrowki 2012).

Howard has a discussion with his friends about Bernadette's opinion on having children and like Alex from *Whitney*, we see that Bernadette's unwillingness to have children is a "deal-breaker." Howard explains to his friends, "I can't see a life where I don't have kids. I mean, people have kids. I mean, even Donkey Kong had Donkey Kong Jr. so he could teach someone how to kidnap princesses and throw barrels at Italian plumbers" (Lorre, Holland, Hernandez and Cendrowki 2012). When Howard and Bernadette see each other again, Bernadette presents a solution to Howard—one that allows for her having children but with a caveat.

BERNADETTE:	Look, it's obvious having kids is really important to you, and I think I came up with a solution.
HOWARD:	Really? That's great. What?
BERNADETTE:	Well, seeing as how I make way more money than you anyway, what if I work and you stay home with the kids?
HOWARD:	Me?
BERNADETTE:	Yeah, you know, you'll watch *Barney* and pull cheerios out of their noses and go on playdates and I'll work and have conversations with people my own age and enjoy my life.
HOWARD:	Yeah.
BERNADETTE:	Yay, so we're good?
HOWARD:	Yeah, we're good.
	(Lorre, Holland, Hernandez and Cendrowki 2012).

By coming up with a solution—one where she agrees to have children but not have to be the one solely responsible for raising them—Bernadette is able to preserve her romantic relationship with Howard that ultimately leads to marriage. However, the conversations about children do not stop once Howard and Bernadette get married. In the episode, "The Spock Resonance," Howard discovers that Bernadette told her father that he is the one that does not want to have children. Bernadette's father, Mike, mentions to Howard the possibility of converting a room into a nursery but adds "I know you don't want to hear this." Howard asks, "Why wouldn't I want to hear it?" Mike responds, "Because Bernie said you didn't want kids." Howard explains that the information is incorrect and that Bernadette is the one who does not want children. As Mike and Howard stand under a building, Bernadette begins stomping on the floor above them yelling "Earthquake!" to get them to stop the conversation (Lorre, Reynolds, Hernandez and Lorre 2015). At the end of the episode, Bernadette and Howard have another conversation about children. Bernadette explains that she lied to her father because she "didn't want to disappoint him" (Lorre, Reynolds, Hernandez and Lorre 2015). The idea that she would disappoint her father because she does not want to have children further exemplifies the idea that choosing to be childfree is not normal and, for the most part, unacceptable. The conversation continues,

HOWARD:	Alright, I admit that I don't help out a lot and I need to work on that but if we had a kid, it would be different.
BERNADETTE:	Why?
HOWARD:	Because when my dad left, I promised myself that if I ever had a chance to be a father, I'd always be there.
BERNADETTE:	Okay, I'll think about it.
HOWARD:	And this is an actual 'I'll think about it,' not like the 'can we get a motorcycle with a sidecar' I'll think about it?
BERNADETTE:	It's an actual 'I'll think about it.'
HOWARD:	I really believe I'd be a great dad.
BERNADETTE:	I know you would.
	(Lorre, Reynolds, Hernandez and Lorre 2015).

While Whitney and Bernadette have to choose the path of children in order to maintain a their romantic relationships, popular culture has provided us with at least one example of a television character choosing not to have children in order to stay in a romantic relationship. Carrie Bradshaw, the main character in *Sex and the City*, debates the need to have children towards the end of the series. Throughout the series, audiences see Carrie address the need to value women's choices—whether it is to remain single or childfree. For instance, in the episode "A Woman's Right to Shoes," Carrie is required to remove her shoes upon entry to a baby shower at her friend Kira's house (Bicks and Van Patten 2003). Upon leaving the shower, she discovers that her shoes, a pair of Manolo Blahniks worth over $400, were taken. Upon this discovery, she approaches Kira and is quickly given a pair of outdated sneakers in which to walk home. Carrie later returns to Kira's apartment to see if her shoes have been found.

KIRA:	I should have offered to pay you for them.
CARRIE:	No, no, you don't have to do that.
KIRA:	You know, you have kids and lose all sense of social decency. [opens door] Come in.
CARRIE:	Okay.
KIRA:	So, um, how much were they?
CARRIE:	$485.
KIRA:	Uh. Come on, Carrie, that's insane.
CARRIE:	Well, that's what they cost.
KIRA:	I'll give you $200.
CARRIE:	Okay, this is an awkward conversation.
KIRA:	Sorry. I just think that's crazy to spend that much on shoes.
CARRIE:	You know how much Manolos are, you used to wear them.
KIRA:	Sure, before I had a real life. But Chuck and I have responsibilities now. Kids, houses. $485? Like, wow.
CARRIE:	I have a real life.
KIRA:	No offense, Carrie, but I really don't think we should have to pay for your extravagant lifestyle. I mean, it was your choice to buy shoes that expensive.
CARRIE:	Yes, but it wasn't my choice to take them off.
KIRA:	They're just shoes.
	(Bicks and Van Patten 2003).

Kira's statement that Carrie does not have a "real life" because she is not married and does not have children is symptomatic of our cultural inclinations towards the hegemonic ideal of motherhood and marriage. Carrie questions Miranda, one of her close friends, "Is it bad my life is filled with shoes and not children?" (Bicks and Van Patten 2003). Carrie ponders, "When we were young, Marlo Thomas sang to us about accepting each other and our differences.

But then we got older and we started singing a different tune. We stopped celebrating each other's life choices and started qualifying them. Is acceptance really such a childish concept or did we have it right all along? When did we stop being free to be you and me?" (Bicks and Van Patten 2003). Carrie makes the decision to register for a pair of shoes for a wedding to herself and Kira promptly purchases the shoes as a gift to Carrie. While the audience is left believing that Carrie is happy with her life decisions, the question remains—is

Carrie missing out on something because she is not married and does not have children?

Carrie finds herself at a point towards the end of the series where she has to explicitly address whether or not she wants to have children. In the early stages of a relationship with Aleksandr (often referred to as "the Russian"), she learns that he has already been married and has a child. In the midst of this discovery, Carrie begins debating whether or not, at the age of 38, her future involves biological children. When asked by Aleksandr if she ever wanted children, Carrie responds, "Ah well, you know, I've always thought that I might. I just, I just haven't gotten around to it yet" (Chupack and Engler 2004). Aleksandr makes it known to Carrie that he is not interested in having more children. Carrie speaks to Charlotte, one of her close friends about the situation, expressing "If I really wanted to have a baby, wouldn't I have tried to have one by now? I wanted to be a writer. I made myself a writer. If I want a ridiculously extravagant pair of shoes, I find a way to buy them." Charlotte tries to get Carrie to reconsider and asks, "Why should you give up having a baby for a man you hardly know?" Carrie responds, "Why should I give up a man for a baby I hardly know I want?" This leads Carrie to question, "Did we *want* babies and perfect honeymoons or did we *think* we *should* want babies and perfect honeymoons?"

As Carrie continues to debate the issue of motherhood, Samantha—her one friend who has never wanted children—states, "There are a lot of fabulous things in life that don't include a baby. What would that be like?" Carrie replies, "Well, him, sex and travel. Comfort. Love and extraordinary adventures."

Samantha exclaims, "Not too shabby." (Chupack and Engler 2004). While Carrie does not provide a definitive answer to the baby question at the end of the episode, the issue is dealt with in regards to Aleksandr. If she stays with him, she will not have children and currently, her life with him is something she does not wish to abandon.

Like Whitney and Bernadette, Carrie needs to make a decision about her romantic relationship. Unlike Whitney and Bernadette, who adhere to the hegemonic ideal of motherhood and agree to the possibility of having children in the future to keep their romantic relationships stable, Carrie abandons the hegemonic ideal of motherhood in order to maintain her romantic relationship. Interestingly, regardless of the outcome, women are willing to sacrifice their own desires in order to maintain a romantic relationship. This reinforces the heterosexual imaginary or "the belief system that relies on romantic and sacred notions of heterosexuality in order to create and maintain the illusion of well-being" (Ingraham 1999, p. 16). For Whitney, Bernadette, and Carrie, romance and love are the ultimate aims.

Childfree by Choice, Romance-Free by Consequence

While there are several female characters that make decisions regarding their desires to be a mother based upon sustaining romantic relationships, there are a few clear examples of female television characters who stay true to their desires not to have children regardless of how that may impact their romantic relationships. Often writers choose to place these characters in extreme situations in order to make it clear to the audience their desires not to have children.

The first example is Robin Scherbatsky from *How I Met Your Mother. How I Met Your Mother* aired from 2005-2014 and followed a group of friends as one of them, Ted Mosby, explains to his children how he met their mother. In the "Pilot" episode, Ted meets Robin and on their first date he proclaims, "I love you" (Bays, Thomas and Fryman 2005). While Ted makes it clear at the end of the episode to his children that this is how he met Aunt Robin and not his children's mother, the romance between Robin and Ted continues to develop throughout the first season and the two enter into a romantic relationship even though Robin has explicitly stated to Ted that she is uninterested in marriage or having children. Their relationship comes to a breaking point in "Something Blue." Robin says to Ted, "You want to have kids. Ted, you know that kids were never a part of my plan. But if I were going to have someone's babies, I'd have your babies" (Bays, Thomas and Fryman 2007). At this point, Robin is attempting to save the relationship by agreeing to give Ted what he wants. However, after a short period of time where the two discuss having children and moving to Argentina for Robin's work, Ted inquires, "We're not doing this are we?" Robin admits, "I don't want to have kids in Argentina." Ted responds, "And I don't want to have kids in Argentina." Robin admits to Ted that she does not want children even if that means the dissolution of their romantic relationship. In the voiceover to his children at the end of the episode, Ted reflects,

> As hard as it was at the time, we both got what we wanted. She did eventually go on to live in Argentina and Morocco, Greece, Russia, even Japan for a little while. And I, well, I met your mom. I think for the most

> part if you're really honest with yourself about what you want out of life, life gives it to you. (Bays, Thomas and Fryman 2007)

Later in the series, Robin has a pregnancy scare (Kelly and Fryman 2011). The audience is aware that Robin does not want children and she makes that clear to Barney, her friend and the man who would be the father. Robin explains, "This is not good news. I don't want kids. I've never wanted kids and never in a million years will I ever want kids. I can't have a baby." When Robin and Barney go to the doctor's office for an appointment for a pregnancy test, the two break out in a dance when the doctor exclaims, "You're not pregnant." Yet, the storyline does not end there. Robin is called back into the doctor's office and is told that she can never have a child due to infertility. She states, "Great. I mean, not great. I've never wanted to have kids so if it had to happen to someone." Later, Robin becomes melancholy about her choice being taken away from her. Grose (2011) explains that Robin is "mourning the curtailing of possibility." She does not want to tell her friends because she does not want to discuss it so instead when asked what is wrong, she substitutes pole-vaulting for motherhood to Lily, her best friend:

> ROBIN: I can never be a pole-vaulter for the Canadian Olympic Team. I'm too tall or something.
>
> LILY: I don't get it, Robin. Did you really want to be a pole-vaulter?
>
> ROBIN: No. No, I was, uh, I was always adamantly against having a pole-vaulting career even though that's what most women want.
>
> LILY: Most women want to be a pole-vaulter?
>
> ROBIN: In Canada. It's very big up there. You know, it's meet a nice guy, get married, vault some poles, but I never wanted that. Of course, it's one thing not to want something, it's another thing to be told you can't have it. I guess it's just nice knowing that you could someday do it if you changed your mind. But, now all of a sudden that door is closed.
> (Kelly and Fryman 2011)

Robin explains to her imaginary children in a voiceover, "So I can't have kids. Big deal. This way there is no one to hold me back in life. No one to keep me from traveling to where I want to travel. No one to get in the way of my career" (Kelly and Fryman 2011).

While Robin is honest about her feelings about not wanting to have children from the beginning of the series, the writers make it so that Robin cannot change her mind. Her infertility leads to the deterioration of her engagement with her fiancé, Kevin. Robin does have some success with romantic relationships in that she ends up marrying Barney in the final season of the series. However, her marriage to Barney also dissolves as they both realize that what they want is being compromised for their relationship to each other. Barney, the group's playboy, goes back to having sex with random women. However, it is when he finds out that one of the women he had sex with is pregnant that he realizes what he needed all along—a child to change his ways and make him mature. This development puts a fine point on Robin's infertility in that she was never going to be able to give Barney what he really needed in life—a child.

Furthermore, at the end of the series Ted's wife, "the mother," dies and his children encourage him to go and ask out Aunt Robin. The series comes full circle with a scene similar to Ted's first date with Robin in the "Pilot" episode (Bays, Thomas and Fryman 2005) to the final scene in the last episode "Last Forever" (Bays, Thomas and Fryman 2014). From the beginning of the series, Ted admits that he loves Robin. While they have an off-again, on-again relationship that evolves into a friendship, the final scene demonstrates that Ted loved Robin all along. However, in order to get what he really wanted in life—children—he needed to find someone else to have them as Robin did not want children and eventually could not physically have them. Again, this adds emphasis to the importance of children and motherhood in our culture. Because Robin was unwilling (and unable) to have children, her romantic relationships were compromised—at least until children were provided by another person.

All of the examples of childfree women up to this point have been in television sitcoms and the storylines tend to rely on some level of comedy in which to deal with potential challenges to the hegemonic ideal of motherhood. As Cooper and Pease (2002, p.301)) explain, "Social critiques. often are articulated within comic narratives, in which hegemonic ideals can be safely questioned by lampooning them because the humorous context allows audience members to participate in the mockery without feeling threatened or alienated." However, there is a clear example of being childfree by choice—regardless of what that decision may do to romantic relationships—in the television drama, *Grey's Anatomy,* through the character of Cristina Yang.

Grey's Anatomy is a medical drama that follows the progression of a group of medical interns to becoming surgeons. Cristina Yang is one of the most determined in the group to be the best at her job. For Cristina, being a surgeon is her number one priority. She makes this clear to Meredith Grey, a fellow intern, during the episode "Enough is Enough" when she discovers she is pregnant. She says, "You know what happens to pregnant interns. I'm not switching to the vagina squad or spending my life popping zits. I'm too talented. Surgery's my life" (Parriot and Horton 2005). Cristina does not tell the father of the baby, Preston Burke. Instead when she schedules her appointment for an abortion she tells Meredith, "The clinic has a policy. They wouldn't let me confirm my appointment unless I designated an emergency contact person. Someone to be there just in case. To help me home, you know, after. Anyway, I put your name down" (Parriot and Horton 2005). However, before she has a chance to have an abortion, she faints during surgery and it is discovered that she has an ectopic pregnancy (Vernoff and Davidson 2005). Cristina loses the baby and a fallopian tube in the process.

Despite the loss of pregnancy, the relationship between Cristina and Burke continues until their wedding day. When Cristina hesitates walking down the aisle, Burke calls off the wedding realizing that he does not want to make Cristina do something she does not want to do. Cristina admits, "I really think I want this [marriage]" and Burke responds, "And I really wish you didn't think. I really wish you knew" (Phelan, Rater and Corn 2007). With that, Cristina and Burke's romantic relationship ends. Later in the series, Cristina becomes pregnant again—this time to her husband, Owen Hunt. Cristina tells Owen about the pregnancy but she makes it clear to him that she does not want the child and is going to have an abortion.

OWEN: You give in in a marriage. You make sacrifices. I've made many and I'm asking you to do something for me. Can you simply consider doing something for me?

CRISTINA: Have a baby? This isn't pizza versus Thai. You don't give a little on a baby.

OWEN: Cristina, I'm asking you.

CRISTINA: I am saying no.

OWEN: [pauses] Get out.

CRISTINA: Are you, because I'm denying you a baby, you're trying …

OWEN: Because you denied me part of the decision. Because you denied me a marriage.

(Cahn and Corn 2011)

Owen's insistence that Cristina's decision to have an abortion is denying him a marriage falls in line with the hegemonic ideal of motherhood. Even though her feelings on having children have always been clear to Owen, her making the decision to actively end a pregnancy goes against the idea that once you get married, you have children. In contrast to characters like Whitney and Bernadette where we see them compromise on the possibility of having children in order to sustain their romantic relationships, Cristina continues with her choice to remain childfree.

Cristina goes through with the abortion, staying true to her decision not to have children. Owen questions, "Meredith says you didn't do it yet." Cristina responds, "I didn't change my mind." Owen confirms, "I know. You don't change your mind" (Cahn and Phelan 2011). Owen escorts Cristina to her appointment and stays with her through the procedure. However, their relationship deteriorates and reaches a breaking point when Owen exclaims in the episode, "Hope for the Hopeless," "You killed our baby! You don't ever forget that!" (Nowalk and Jackson 2012). Once again, Cristina's romantic relationship deteriorates because of her decision not to become a mother and to focus on her career. Cristina's decision is an anomaly in the representations of women in popular TV programming. For instance, in order to allow for Robin (from *How I met your Mother*) not to have children, the writers make her infertile. "The decision to be a 'NotMom' was made once and for all, but not by her" ("Unfunny Realities of Childfree by Choice" 2014). By making Robin infertile, the writers transform her from a woman who is actively choosing to be childfree to a "sympathy worthy woman" (Graham and Rich 2014, p. 515) because she no longer has a choice. Smith (2013) argues, "In pop culture, childfree people are extremely unusual, and when we do appear, we're usually converted by a pregnancy, a baby suddenly appearing in our lives, or some other outside factor. The message sent by pop culture is that childfree people don't really exist; we just need some persuading to come over to the right side of things." Cristina is an anomaly because she is not persuaded to follow the hegemonic ideal of motherhood even when it means letting go of a romantic relationship. She cannot be converted.

CONCLUSION

Media framing of the hegemonic ideal of motherhood and how this relates to romantic relationships is clear—in the majority of cases, romance trumps all. While there are a few representations of choosing to be childfree even if it means ending up romance-free, the comfortable position is to abide by the hegemonic ideals of romance and love, marriage, and children. For instance, while audience members know that Robin from *How I Met Your Mother* never wanted to have children, making her infertile allows for an acceptance of her decision because individuals who may be uncomfortable with her choosing to be childfree can dismiss it as a circumstance instead of a choice and believe that she might have changed her mind had she met the right person for a romantic relationship. Additionally, Carrie's decision to remain in a romantic relationship with Aleksandr even after he says he does not want to have children is acceptable because she is choosing romance and the potential for a long-term relationship. Unfortunately, media representations of happily childfree women are lacking in television—both in sitcoms and dramas. As a result we are also missing storylines where women choose to remain childfree and are able to have successful romantic relationships regardless of their decision about becoming a mother. The default position of television writers is to bring babies into the mix when someone has entered into a relationship—even if it does not completely make sense. Carlson (2015) questions the decision to bring children into television shows such as *The Mindy Project*, *Parks and Recreation*, and *30 Rock,* writing, "Did the story really need that development, or is that development just expected when it comes to female characters?" In order to follow the life script, babies are the norm. Carlson (2015) goes on to argue,

> As we work as a society to have a more all-encompassing view of what it means to be a parent and what it means to have equal rights, it is equally important to see the kind of variety we preach among our friends presented throughout entertainment. We need more childless women—and more childless couples—depicted in our media. We need to represent those who have decided they don't want children and are happy with it; we need to show that choice is acceptable, and normal.

As the number of women remaining childfree increases in our society, we need to challenge the hegemonic ideals of romance and love, marriage, and children because this is not the life story of everyone. We need variety—romance with children, romance without children, and happiness without romance or children. "Let's recognize that each of us has the capability to be a good human being in our own right, as an individual, whether single or in a relationship, with children or without" (Young 2004, p. 10).

REFERENCES

Bays, C. and Thomas, C. (Writers) and Fryman, P. (Director). (2005). Pilot [Television episode]. C. Bays, C. Thomas, and P. Fryman (Executive Producers), How I Met Your Mother: Season One. Los Angeles, CA: 21st Century Fox.

Bays, C. and Thomas, C. (Writers) and Fryman, P. (Director). (2007). Something blue [Television episode]. C. Bays, P. Fryman, and C. Thomas (Executive Producers), How I Met Your Mother: Season Two. Los Angeles, CA: 21st Century Fox.

Bays, C. and Thomas, C. (Writers) and Fryman, P. (Director). (2014). Last forever [Television episode]. C. Bays, P. Fryman, and C. Thomas (Executive Producers), How I Met Your Mother: Season Nine. Los Angeles, CA: 21st Century Fox.

Bicks, J. (Writer) and Van Patten, T. (Director). (2003). A woman's right to shoes [Television episode]. E. Zuritsky (Executive Producer), Sex and the City Season Six. New York City, NY: HBO Home Studios.

Britt, E. (2001). Conceiving normalcy: Rhetoric, law, and the double binds of infertility. Tuscaloosa, AL: University of Alabama Press.

Cahn, D. (Writer) and Corn, R. (Director). (2011). Unaccompanied minor [Television episode]. S. Rhimes (Executive Producer), Grey's Anatomy: Season Seven. Burbank, CA: Walt Disney Studios.

Cahn, D. (Writer) and Phelan, T. (Director). (2011). She's gone [Television episode]. S. Rhimes (Executive Producer), Grey's Anatomy: Season Eight. Burbank, CA: Walt Disney Studios.

Cain, M. (2001). The childless revolution: What it means to be childless today. Cambridge, MA: Perseus Publishing.

Carlson, S. (2015, March 6). We need more childless women, couples on TV. Retrieved October 15, 2015, from Pajiba.

Chupack, C. (Writer) and Engler, M. (Director). (2004). Catch 38 [Television episode]. E. Zuritsky (Executive Producer), Sex and the City Season Six. New York City, NY: HBO Home Studios.

Cooper, B. and. Pease, E. C. (2002). "Don't want no short people 'round here'": Confronting heterosexism's intolerance through comic and disruptive narratives in Ally McBeal. *Western Journal of Communication*, 300-318.

Corcoran, F. (1984). Television as ideological apparatus: The power and the pleasure. *Critical studies in mass communication*, 131-145.

Cummings, W. (Writer) and Ackerman, A. (Director). (2011a). First date [Television episode]. W. Cummings (Executive Producer), Whitney: Season One. Universal City, CA: Universal Studios, CA.

Cummings, W. (Writer) and Ackerman, A. (Director). (2011b). Spicing things up [Television episode]. W. Cummings (Executive Producer), Whitney: Season One. Universal City, CA: Universal Studios.

Cummings, W. (Writer) and Thomas, B. (Director). (2012). Mad woman [Television episode]. W. Cummings (Executive Producer), Whitney: Season One. Universal City, CA: Universal Studios.

Cummings, W. and Quaintance, J. (Writers) and Ackerman, A. (Director). (2011). Up all night [Television episode]. W. Cummings (Executive Producer), Whitney: Season One. Universal City, CA: Universal Studios.

Davies, B. (1992). Women's subjectivity and feminist stories. In C. Ellis (Ed.), Investigating subjectivity: Research on lived experience (pp. 53-76). Thousand Oaks, CA: Sage.

Entman, R. M. (1993). Framing: Toward clarification of a fractured paradigm. *Journal of Communication*, 51-58.

Galician, M.-L. (2004). Sex, love, and romance in the mass media: Analysis and criticism of unrealistic portrayals and their influence. Mahwah, NJ: Lawrence Erlbaum.

Graham, M. a. (2014). Representations of childless women in the Australian print media. *Feminist Media Studies*, 500-518.

Grose, J. (2011, December 8). Child-free on TV. Retrieved from Slate: http://www.slate.com/articles/double_x/doublex/2011/12/two_sitcoms_pose_the_question_why_aren_t_more_women_in_prime_time_babyless_and_proud_of_it_.html.

Hayden, S. (2011). Constituting savvy aunties: From childless women to child-focused consumers. *Women's Studies in Communication*, 1-19.

Illouz, E. (1997). Consuming the romantic utopia: Love and cultural contradictions of capitalism. Berkeley, CA: University of California Press.

Ingraham, C. (1999). White weddings; Romancing heterosexuality in popular culture. New York: Routledge.

Kelley, J. (Writer) and Fryman, P. (Director). (2011). Symphony of illumination [Television episode]. C. Bays, C. Thomas, and P. Fryman (Executive Producers), How I Met Your Mother: Season Seven. Los Angeles, CA: 21st Century Fox.

Livingston, G. (2010, June 25). Childlessness up among all women; Down among women with advanced degrees. Retrieved from Pew Research

Center: http://www.pewsocialtrends.org/2010/06/25/childlessness-up-among-all-women-down-among-women-with-advanced-degrees/.

Lorre, C., Holland, S., and Hernandez, T. (Writers) and Cendrowski, M. (Director). (2012). The shiny trinket maneuver [Television episode]. C. Lorre and B. Prady (Executive Producers), The Big Bang Theory: Season Five. Burbank, CA: Warner Brothers.

Lorre, C., Reynolds, J. and Hernandez, T. (Writers) and Lorre, N. (Director). (2015). The Spock resonance [Television episode]. C. Lorre and B. Prady (Executive Producers), The Big Bang Theory: Season Nine. Burbank, CA: Warner Brothers.

Lull, J. (2015). Hegemony. In G. Dines (Ed.), Gender, race, and class in media: A critical reader (pp. 39-42). Los Angeles: Sage.

Mantel, H. (2013). No kidding: Women writers on bypassing parenthood. Berkeley, CA: Seal Press.

May, E. T. (1995). Barren in the promised land: Childless Americans and the pursuit of happiness. Cambridge, MA: Harvard University Press.

Minsky, T. (Writer) and Seidelman, S. (Director). (1998). The baby shower [Television episode]. Sex and the City: Season One. New York City, NY: HBO Home Studios.

Notkin, M. (2014). Otherhood: Modern women finding a new kind of happiness. Berkeley, CA: Seal Press.

Nowalk, P. (Writer) and Jackson, M. (Director). (2012). Hope for the hopeless [Television episode]. S. Rhimes (Executive Producer), Grey's Anatomy: Season Eight. Burbank, CA: Walt Disney Studios.

Parriot, J. (Writer) and P. Horton (Director). (2005). Enough is enough [Television episode]. S. Rhimes (Executive Producer), Grey's Anatomy: Season Two. Burbank, CA: Walt Disney Studios.

Phelan, T. and Rater, J. (Writers) and Corn, R. (Director). (2007). Didn't we almost have it all [Television episode]. S. Rhimes (Executive Producer), Grey's Anatomy: Season Three. Burbank, CA: Walt Disney Studios.

Sandler, L. (2013, August 12). The childfree life: When having it all means not having children. Retrieved from Time Magazine: http://time.com/241/having-it-al-without-having-children/.

Sklaver, S. (Writer) and Cendrowski, M. (Director). (2011). Clarence [Television episode]. W. Cummings (Executive Producer), Whitney: Season One. Universal City, CA: Universal Studios. Unfunny realities of childfree by choice AND by chance. (2014, March 31). Retrieved October 15, 2015, from TheNotMom.com: http://thenotmom.com/how-i-met-your-mother-unfunny-realities-of-child-free-by-choice-and-by-chance/.

Vernoff, K. (Writer) and Davidson, A. (Director). (2005). Make me lose control [Television episode]. S. Rhimes (Executive Producer), Grey's Anatomy: Season Two. Burbank, CA: Walt Disney Studios. Whitney. (2015, October 15). Retrieved from IMDB.com: http://www.imdb.com/title/tt1850458/.

Young, A. (2004, July 20). Childless and happy. The Advocate, p. 10.

In: Television and Romance
Editor: Amir Hetsroni
ISBN: 978-1-63485-077-3

Chapter 4

IS BEING A SINGLE MAN OKAY? A CRITICAL ANALYSIS OF MASCULINITY AND SINGLEDOM IN THE SERIES HOW I MET YOUR MOTHER

Jennifer Whalen
University of South Florida, Tampa, FL, US

ABSTRACT

This chapter examines the ways in which media representations of 0interpersonal performances on a prime-time television show speak to and reify the cultural narratives surrounding the stigma of male singledom in contemporary American culture. By focusing on the character of *Barney Stinson* in the series *How I Met Your Mother* the work analyzes the dialectical pull between independence and companionship and problematizes the bachelor performance as a co-dependent childlike enactment.

INTRODUCTION

Through a textual analysis of *How I Met Your Mother* this chapter indicates ways in which media representations of interpersonal performances on prime-time television shows speak to and reify the cultural narratives surrounding the stigma of male singledom in contemporary American culture. In particular, my analysis concentrates on one character, *Barney Stinton,* who is a perennial bachelor and womanizer, and on the differences and similarities between him and other protagonists who are more committed (in theory or in practice) to long-term relationships. I bring examples from the first five seasons of the series which indicate ways in which the show reifies the cultural narrative of "singleness as problematic." Concepts that are used throughout the analysis include hegemonic and hyper masculinity, singledom, front-stage and backstage performances.

THEORETICAL UNDERPINNINGS

The term hegemonic masculinity, which is very popular in critical media research, denotes "the general… assumptions and beliefs about masculinity that have become common sense" (Hanke 1998. p. 185). Hegemonic outlooks are taken-for-granted embodiments that are internalized and then reified through social structures such as mass media. This notion of hegemonic masculinity has been critically analyzed in various popular shows such as *The A-Team* (Schwichtenberg 1987), *Miami Vice* (Fiske 1987) and *thirtysomething* (Hanke 1990). More recently, Hatfield (2010) analyzed the sitcom *Two and a Half Men*. She used narrative thematic analysis to pinpoint that media representations of hegemonic masculinity portray what it means "to be a man." While Hatfield's work showed how hegemonic masculinity is reinforced, my analysis examines how men's gender performances play a role in the enactment of romantic relationships, and thus speak to a larger cultural narrative of singledom.

Without undermining biological differences between males and females, we have to remember that gender is also a socially constructed practice in time and space that serves as an ongoing reproduction, reifying stereotypes and behaviors of men and women (Wood 2005). When it comes to gendered performances, Goffman (1967) offers insight on front-stage and backstage performances of personality which is important when examining how characters in a television show perform their role. The audience is provided with insight on both the front-stage and backstage performances of a protagonist—that is, while viewing the show the audience sees the character act in both public and private spaces. Public spaces may be places like bars or diners, while private spaces are either intrapersonal thoughts and feelings or interpersonal interactions with other characters; however, private moments can sometimes occur in a public space. These spaces offer an insight as to the characters' "personality" because they provide a context wherein performances are socially appropriate in a particular situation, but they also demonstrate the backstage performance of characters, which typically reveals more private information about their value system(s). It is important to analyze the context within which gender performances are occurring when doing a critical analysis of a television show because what is being said, and how, and to whom, and in what situation, are all reflections of ways in which gendered patterns of interaction are occurring, therefore, speaking to an embodied and ongoing cultural narrative.

Barney Stinton from *How I met your Mother* is an interesting character to explore because of his brash comments about women and sex. His character is played by Neil Patrick Harris, an openly gay man. While not all audience members are necessarily aware of Harris' sexual preferences, the fact that Harris plays a hypermasculine role of a womanizer makes the character of Barney a unique site of inquiry because there is an irony to the casting team's selection of a gay man as womanizer.

Much of the critical scholarship exploring media portrayals of men on prime-time television shows focuses on the sexual and non-sexual violence that men demonstrate (Cuklanz 1998). While these texts are useful in critical analysis of television portrayals of male characters, the goal of this project is to look at how Barney speaks to a cultural narrative of male singledom, rather than focusing on the objectification of women he displays in his behaviors. His actions may be covertly aggressive, but I want to look at how his behaviors speak to the larger cultural narrative of singledom, which is a topic that hitherto has not yet been extensively explored in prime-time television research. There is some rather outdated literature on the portrayal of single women in prime-time television programming (see Signorielli 1982), but hardly anything has been written about single male characters. This gap in research calls for an analysis of men's singledom, how singleness is perpetuated and reified through prime-time television shows, and the ways in which the characters speak to a larger cultural narrative about masculinity and singleness.

ANALYZING *HOW I MET YOUR MOTHER*

How I Met Your Mother is an award winning prime-time sitcom which was broadcast for nine seasons (2005-2014) in the United States and imported successfully to dozens of countries. Local adaptations also exist (e.g., a Russian version). The original series follows the lives of five friends—Ted, Marshall, Lily, Barney, and Robin—living in the Big Apple and the stories of their failures and successes in friendship and romantic relationships. All of the characters in the show are white, middle class, college-educated, and are in their late twenties in the beginning of the series. Ted, the main character (and the narrator), is a hopeless romantic who is continuously on a quest during the series to find "the one." Barney is a womanizer, constantly in quest of his next sexual partner, and takes great stock in ensuring that he stays out of long-term relationships throughout most of the show. His personality is overtly sociopathic, yet there are moments where Barney comes across as caring and loveable, which redeems him for some of his unsightly relational behaviors. He is also written to be witty and funny, which provides an appeal to his character. The fact that he is played by an openly gay actor adds to the irony.

The series is set in New York City and most of the scenes are shot at a pub, McLarens, the common meeting place for the group. The producers of the show utilize McLarens as a site for Barney to pick up women and discuss his sexual endeavors with his friends. McLarens is also conveniently located next to Ted, Marshall, and Lily's apartment,. Since these characters are "regulars," they often know they can find each other at the bar, which is convenient in showing the characters together in order to perpetuate the overarching theme of the quest to avoid singledom.

The voiceover narration that Ted does throughout the series is framed as being directed to his two children, who are seated on a couch and seemingly bored with the story. It is implied throughout the show by a repeated flashback to the children that how Ted met his wife is a long and complicated story, perpetuating the cultural narrative that a continuous pursuit of a romantic relationship will ensure companionship, as opposed to singledom. The narrative is told primarily through clips of relationships he has with friends, as well as flashbacks that help set the scene for the explanation of his relational failings that helped him find and marry his wife.

While Ted's character also performs the cultural narrative of the importance of finding companionship, the character of Barney Stinson was specifically chosen as the focus of this work because of his performance of the bachelor role and his seeming desire to enjoy living the bachelor lifestyle. While there may be a cultural narrative of singledom that runs across various television programs, what is particularly interesting about Barney's character in this show is his front-stage performances of hypermasculine singledom when his backstage performances demonstrate a need for companionship.

These behaviors pull at the dialectical tension between autonomy and connection (see Baxter 1988), and provide for an interesting look into cultural narratives of male singleness.

Barney's behaviors in the show often involve objectifying women. For example, in the fifth episode of the third season he says: "People at work have this very strange expectation that I, well, work. I know, I know. So to make my colleagues think I'm busting my hump while I'm really out humping busty chicks." Barney is also known to challenge himself in terms of his sexual endeavors. There is an entire episode devoted to Barney's attempt to pull off "The Perfect Week," in which he has sex with seven women in seven days (Season 5, Episode 14). At another occasion he gloats about the number of women he has slept with, which, by the end of the episode mount to 200 (Season 4, Episode 22).

This seemingly problematic bachelor status is introduced within the first few scenes of the Pilot episode of *How I Met Your Mother*. In the voiceover of the very first scene of the series Ted explains that he was 27 years old and was successful and happy at his new job, until his best friend Marshall "screwed things up." The scene cuts to Marshall who is down on one knee in front of Ted, practicing his proposal to his long-time girlfriend, Lily. Ted feels excited for Marshall, but after Marshall rehearses the proposal he asks Ted what his plans are for the night. Ted rolls his eyes and the scene flashes to Barney Stinson, who is on the phone with Ted. Barney begins talking about how he has a new favorite type of woman that he likes. It is implied that the phone conversation had taken place earlier in the day, and that Ted is reflecting back on his plans for the evening. During the phone conversation scene, Barney is at a barbershop, wearing a suit and being shaved by the barber, and is smug in his communication with Ted about his new favorite type of woman, implying that he has had multiple "types" of women that he has been interested in. Ted and Barney then agree over the phone to meet up at McLarens. In the scene following the phone conversation, Ted and Barney are standing at the bar and, Ted, with disgust and concern, expresses to Barney that, "it's always been Marshall, Lily, and me, but now it will be Marshall and Lily… and me. They'll get married and start a family and before long I'm that weird middle-aged bachelor their kids call 'Uncle Ted.'"

This scene exposes the viewers to the idea that a middle-aged bachelor is an unattractive concept. After Ted expresses his fear and concern of being a single man and a third wheel to two of his friends, Barney slaps Ted and asks him if he remembers the first thing that he ever said to Ted when they met. The scene cuts to Barney sliding into a booth next to Ted, who is at McLarens, and he begins explaining that Ted has "not yet lived" because he has not led a life in pursuit of sleeping with lots of women.

This pilot episode sets up the characters' personalities and relationships to one another, but also helps in explaining where the characters are in their romantic lives. Ted's disgust for having to go out to a bar while Marshall will be getting engaged and his conversation with Barney at McLarens emphasize that he does not want to be single, but rather, that he would prefer to settle down and find a woman to commit to. Barney, on the other hand, tries to sell Ted the idea that being in a relationship is overrated, and that he is not really "living" if he is not sleeping with many different women. As Lindsey (1997) notes, contemporary American culture allows men to put an "emphasis on sexual prowess, sexual conquest, and sexual aggression" (p. 225). Barney enacts a hypermasculine approach to singledom that continues throughout the course of the series.

DePaulo (2006) addresses the idea that when a person is single it is, "all about you," and that, "like a child, you are self-centered and immature and your time isn't worth anything since you have nothing to do but play" (Ch. 6 - subtitle). Barney perpetuates this notion in his attempt to convince Ted that being a bachelor is like a playground. He continues throughout the course of the show to reify the idea that the bachelor status is better than being in a romantic relationship, which he often does through hyper-sexualized communication practices regarding women. Here is an example from Second 2, Episode 2:

Barney:	This is the moment I've been waiting for. Starting tonight, I am gonna teach you how to live. Ted, you had your chance. You're out, Marshall's in.
Ted:	Yesss!
Marshall:	Oh god...
Barney:	Marshall, being a single guy in New York City is like ... What's something everybody likes?
Marshall:	Candy...
Barney:	Yes! It's like being in a candy store! You just walk right in and grab yourself some Whoppers!

Barney Stinson is a single man who can be perceived as "playing the game," especially because his attempts to sleep with women are recorded in, what he refers to as, his "playbook," however, his use of props and magic in order to sleep with women, as well as his perceptions of sexual activity as "playground behaviors" situate him also in a childlike position, reaffirming DePaulo's (2006) idea about singledom as self-centered play.

Several episodes into the series, the viewer finds out that years before the timeframe when the series has begun Barney had worn his heart on his sleeve in a long-term relationship, singing romantic songs to his sweetheart and writing her poetry, but she left him for a masculine man who always wore a suit. Because of that failed relationship, Barney has become a chronic womanizer, and is always dressed in a suit, as he thinks that performing hyper-masculinity is the best way to get women. In the fifteenth episode of the first season Barney is shown sobbing over the breakup with his ex-girlfriend whom he lost for the man in the suit. In an intense scene, he "suits up" for the first time, cuts off his long ponytail, shaves his face, and puts on a jacket and a tie. When the process is complete, he winks at the camera. His actions speak to Butler's (1993) notion that "real men pursue sex" (p. 146). Once transformed from a more feminine man into a clean-cut, "suited up" man, Barney finds that he is able to have the confidence because of his physical appearance to pursue, and achieve, sexual conquests.

Interestingly, however, Barney repeatedly communicates to his male friends that he does not want to be the only bachelor and that he needs a reliable "wingman" to go along with him on his sexual conquests. This mediated notion that the bachelor does not want to "play" the game alone perpetuates the notion that men are lonely without a woman, but also that independence, in itself, is uncomfortable. In one of the episodes, Ted, Barney, and Marshall are all single at the same time, and Barney expresses his satisfaction that he has his male friends to pursue women with:

> "For the first time...ever...the three of us are single at the same time. I've dreamed about this day, boys, and it is going to be le...gen...dary! Together we will own this city. Any time a girl wants to get back at her ex-boyfriend, we'll be there. Any time a girl wants to...solve her father issues through promiscuity and binge drinking, we will be there. Any time a bachelorette party drives through the city in a limo, sticking their heads out the sunroof, shouting "What's up New York!" we will be what is "up" New York!."

While Barney continues to express his lifestyle as "awesome" and his sexual pursuits as "legendary," the series shows that he actually prefers to be in company with other people through his friendships and sexual activity. Perlman and Peplau (1981) note that one way individuals cope with loneliness is to achieve higher levels of social contact. This translates into attempting to seek out relationships and lowering the standards set for acceptable companions.

Indeed, after Barney's breakup and subsequent makeover, he begins to accept nearly any available woman as an appropriate partner. This typifies the bachelor's need of attention.

It is also interesting that Barney is portrayed as a character with further emotional flaws as a result of his early childhood, as he grew up being raised by a single mom who told him that his father was the famous television character, *Bob Barker*.

In contrast, the other characters in the show are portrayed as having been raised by a functioning nuclear family. While these other characters make peace with their sometimes estranged parents, Barney struggles throughout the series to find stability in the absence of his father, even going so far as to believe that Bob Barker is really his dad, which causes him to compete in *Jeopardy* just to meet Bob Barker.

In addition to the woman that left him and caused him to change his appearance and his attitude toward women, Barney's familial background provides an excuse for him to avoid long-term romantic relationships. This also gives the viewer an opportunity to be sympathetic toward Barney because of his problematic family background.

Research conducted on loneliness indicates that, while it is a common experience, it is also a distressing feeling (Perlman and Peplau 1981). Barney acts as an "expert" on attracting women, continuously helping his single friends to try and find females to sleep with.

This suggests that Barney finds it problematic that his friends do not have the companionship they are supposedly seeking, which, again, speaks to a larger cultural narrative of loneliness as problematic.

While Barney repeatedly expresses a desire for his male friends to be single, so that he can have good time with them, there are many moments in the series where he still pushes for Ted and Marshall to be in romantic relationships, acting as though he wants to be alone, but demonstrating that he still has a stake in love relationships. He fights for his friends in order to prevent them from forfeiting their chances at love and marriage, and even fights for them, especially Lily and Marshall, to stay together after they have an emotional breakup. He goes so far as flying across the country to try and convince Lily to get back together with Marshall. However, he makes Lily swear not to tell anyone that he went to see her to have that conversation in order to maintain his front-stage performance of a womanizer who distrusts long-term relationships.

Conclusion

In this chapter, I have argued that the character of *Barney Stinson* in the series *How I Met Your Mother* enacts hegemonic masculinity and stereotypical gendered performances in front-stage and backstage performances of his personality, reifying the notion that being a bachelor is distasteful and problematic. I have analyzed how the bachelor performs as a child, desperate for attention and struggling with lack of companionship. I have also focused on the ways in which the bachelor lifestyle problematizes the notion of singledom and perpetuates the idea that being single equates to being lonely. This demonstrated the dialectical presentation of single men on prime-time TV programming.

References

Baxter, L. (1988). A dialectical perspective on communication strategies in relationship development. In S. W. Duck, D. F. Hay, S. E. Hobfoll, W. Iches, and B. Montgomery (Eds.) Handbook of personal relationships (pp. 257-273). London: Wiley.

Butler, J. (1993). Bodies that matter: On the discursive limits of sex. New York: Routledge.

Cuklanz, L. M. (1998). The masculine ideal: Rape on prime-time television, 1976–1978. *Critical Studies in Media Communication*, 15(4), 423-448.

DePaulo, B. M. (2006). Singled out: How singles are stereotyped, stigmatized, and ignored and still live happily ever after. London: St. Martin's Griffin.

Fiske, J. (2010). Television culture. London: Taylor and Francis.

Goffman, E. (1967). On face-work. *Interaction ritual*, 5-45.

Hanke, R. (1990). Hegemonic masculinity in thirty something. *Critical Studies in Media Communication*, 7(3), 231-248.

Hanke, R. (1998). Theorizing masculinity with/in the media. *Communication Theory*, 8(2), 183-201.

Hatfield, E. F. (2010). "What it Means to Be a Man": Examining Hegemonic Masculinity in Two and a Half Men. *Communication, Culture and Critique*, 3(4), 526-548.

Lindsey, L. (1997) Gender roles: A sociological perspective. Upper Saddle River, NJ: Prentice Hall.

Perlman, D., and Peplau, L. A. (1981). Toward a social psychology of loneliness. *Personal relationships*, 3, 31-56.

Schwichtenberg, C. (1987). Articulating the people's politics: manhood and right-wing populism in *The A-team Communication*, 9(3-4), 379-398.

Signorielli, N. (1982). Marital status in television drama: A case of reduced options. *Journal of Broadcasting and Electronic Media*, 26(2), 585-597.

Wood, J. T. (2005). Gendered lives: Communication, gender, and culture. Belmont, CA: Wadsworth Publishing.

SECTION II: EMPIRICAL INVESTIGATIONS

In: Television and Romance
Editor: Amir Hetsroni
ISBN: 978-1-63485-077-3

Chapter 5

Sex and the Senior: An Analysis of Older Adults' Sexual Activity Portrayed in Popular US Television

Heather L. Hundley and Jo Anna Grant
California State University, San Bernardino, CA, US

Abstract

Senior sex is exceedingly rare on US prime time television shows. Using content analysis and critical discourse analysis, this study examines 40 episodes from 6 series of popular US prime time television aired from 2009-2014 which contain portrayals of senior citizens engaging in sex. Results show that senior sex is between single heterosexuals, close to the same age, and from the same race. Additionally, the seniors complement each other's sexual performance, do not display signs of jealousy, and maintain a range of relational commitment (from one-night stands to long term marital relationships). Senior sexual relationships are predominantly public, and while friends are supportive, family members are not. These results are compared with social beliefs and stereotypes of senior sex, population data on senior sex, and sexual portrayals on TV and other media.

Introduction

Older adults are commonly thought to be asexual. They are often assumed to be uninterested in sexual activity or not capable of it. Even worse, older adult sexual activity is thought to be "dirty" (Schultz, Noelker, Rockwood, & Sprott, 2006, p. 1064) or taboo (Woodward, 1991). This belief is reflected in two prominent stereotypes of sexually active older adults: the "dirty old man" and the predatory "cougar" for women.

However, these negative stereotypes of older adults' sexuality do not reflect reality. Nussbaum, Miller-Day and Fisher (2009) interviewed older adults and concluded that, contrary to these negative cultural stereotypes of old age, "pleasurable sexual encounters are the norm in older adulthood" (p. 47). More specifically, 73% of 57-64 year olds, 53% of 65-75 year olds, and 26% of 75-85 year olds said they had sex with their partner in the last year, and among sexually active seniors, most reported having sex at least 2-3 times per month (Lindau, Schumm, Laumann, Levinson, O'Muircheartaigh, & Wait, 2007). A more recent study based on the *English Longitudinal Study of Health* echoes this finding reporting that 54% of men and 34% of women in their 70's and 80's disclose having sex at least twice a month (Lee, Nazroo, O'Connor, Blake, & Pendelton, 2015, January 27). For many older adults, especially older women, the greatest limiting factor to sexual activity is actually the availability of a partner, not the lack of sexual desire or ability (Karraker, DeLamater, & Schwartz, 2011). This chapter employed *content analysis* and *critical discourse analysis* to examine the portrayal of senior citizens engaging in sex in popular US prime time television that aired during the first half of the 2010s.

Media Representations of Older Adults Sex and Romance

Despite their inaccuracy, these negative stereotypes continue into media representations of older adults and their sexuality. In general, media depictions of older adults are sparse and portray them negatively as marginal to society, dependent, mean and bitter (see Fealy, McNamara, Treacy, & Lyons, 2011; McContha, Schnell, & McKenna, 1999; Robinson, Skill, & Turner, 2004; Zhang, Harwood, Williams, Ylänne-McEwen, Wadleigh, & Thimm, 2006). However, depictions of older adults in romantic and sexual relationships are even more infrequent. Such depictions play off of existing stereotypes by either marginalizing older adults' sexuality, portraying older adult sexual activity in subtly suggestive and metaphorical ways, or hyper-sexualizing older adults for the shock or comedic value (Palmer, 1994).

Seniors in Advertising

The most common media portrayal of older adults occurs in advertisements (Zhang, Harwood, Williams, Ylänne-McEwen, Wadleigh, & Thimm, 2006). Yet, even advertisements targeted at older adults perpetuate negative stereotypes of sexual desirability. Baker and Goggin (1994) conducted a content analysis of the advertisements in *Modern Maturity*, a magazine published by the American Association of Retired Persons. They found that the latent message was that the natural effects of aging reduce sexual attractiveness and sexual intimacy. The products advertised included non-stick chewing gum, bladder control products, and denture cleaners and promised quick solutions to gum sticking to dentures, weak bladders, and unhygienic dentures respectively. One product declared: "*So go ahead—with Freedent, you can get close with confidence*," another affirmed "*With Serenity, I'm not afraid to get close anymore*" (p. 143). Thus, while older adults may desire personal intimacy, they are inhibited to do so unless they purchase and use specific products allowing them to become more socially acceptable.

In contrast to the negative implications of sexual desirability in these advertisements, Williams, Ylanne, and Wadleigh's (2007) case study of the Olivio/Bertolli margarine print campaign from 1996-2003 found more positive depictions of aging and sexuality. They found that the campaign went through four phases each with different and positive depictions of older adults. The second phase slightly pushed the boundaries of older adult stereotypes by gently implying—not depicting—older adults' sexual relationships. The text of one ad, with no people portrayed stated:

> A seventy-year-old couple demonstrating the benefits of a Mediterranean diet. There's probably not much call for aphrodisiacs in the Mediterranean… which has got to be good for your family's health, not to mention the… *ahem*… active part you and your partner will take later in life. (Williams, et al., 2007, p. 12)

The final phase of the Olivio/Bertolli campaign pushed the boundaries even further by actually foregrounding older adult romance and sexual attraction by depicting older adults in attractive and romantic poses and marking them with terms such as “babe magnet,” and “arm candy” in the text of the ads. Yet even these ads marginally suggested sexual intimacy. Furthermore, the ad which was the most sexually suggestive made older adults literally invisible.

Ads for Viagra and other erectile dysfunction (ED) medication ads evoke older adult sexuality, yet they also use innuendo to hint at the topic rather than explicitly address it, and frequently avoid showing older adults. Irvine (2006) traced the Viagra advertising campaign and argues that beginning with Bob Dole (then 75 years old) as a spokesman, the ad campaign was designed to give older men courage to speak to their doctors about impotence, reinforcing stereotypes about older adults unable to be sexual and offering a medical solution. However, once sales flattened in 2000, Pfizer and other ED drug companies abandoned images of older adults and started featuring men in their 30s and 40s, raising Viagra sales 312% for men aged 18-45. Even with middle aged adults, the ads used metaphor to imply sex rather than discuss it directly. For example, one Levitra ad featured a 30-something year old male attempting to throw a football through a tire swing, only making the shot through the hole after using the product.

As with the Olivio/Bertolli ads (Williams, et al., 2007), the ED drug campaigns became more explicit as they continued, but still avoided the bare skin and bulging genitalia commonly seen in advertisements for blue jeans and underwear, and instead relied on sexual innuendo (Irvine, 2006). Viagra used dance as a metaphor for sex in its “Let the Dance Begin” campaign. Cialis implied, but did not show skin with its now trademark his and hers bathtubs unusually placed in a grassy meadow. Sometimes the bathtubs were occupied with older adults looking out into a beautiful landscape and sometimes they were empty, implying an affirmative answer to Cialis’ slogan question “Will You Be Ready?” Cialis’ marketers noted that they specifically chose separate bathtubs rather than a single hot tub because “we don’t want to put Cialis that close to the sex act” (Irvine, 2006, p. 42). These ads reinforce Woodward’s (1991) assertion that senior sex is inherently “unwatchable.”

Another genre of advertisement that more typically identifies sexual intimacy is personal ads. Once seen more often in newspapers, personal ads have created a new marketplace on the Internet. Some examples include Senior People Meet, 50plus-Club, and Single and Over 50. This is particularly interesting because it consists of self-portrayals of aging rather than externally constructed ones.

Coupland (2000) examined personal ads to discover the relational goals older adults revealed in them. She found that older adults "articulate less ambitious, more modest, and certainly less sexual relational goals" (p. 28) than younger users of personal ads. Older adults' goals tended to be more social and recreational than intimate. As is typical of the medium, users were inclined to self-present in a positive manner, however, older adults used different strategies to do so. They were apt to pay less attention to attractiveness and physicality by not foregrounding these features either of themselves or of their sought after mate. Older adults also used various forms of age denial to build a positive presentation, such as creating a disjunction between the age in years and the perceived age (e.g., "look 40's"). These presentations suggest that, while the advertiser was old, he/she was not "typically" so.

Alterovitz and Mendelsohn (2013) examined the personal ads of middle-aged (40-54), young-old (60-74), and old-old (75+) adults on Internet dating sites seeking heterosexual relationships. The following categories emerged: loneliness, adventure/exploration, seeking soul mate, romantic activities, sexual activities, and health. Across the categories, little difference appeared between the middle-aged and young-old groups. However, 5 of the 6 categories showed a significant difference between the middle-aged and old-old. Middle-agers were more likely to mention adventure/exploration, seeking soul mate, romantic activities, sexual activities, while the old-old were more likely to mention their health status. The old-old and the young-old were equally likely to mention loneliness in their ads.

Personal and commercial advertisements portraying older adults have several characteristics in common. They tend to divert attention away from the age of older adults by reframing age. This is done visually by using younger actors as in the ED campaigns (Irvine, 2006), portraying older adults engaging in strenuous activities like yoga and soccer (Williams, et al. 2007), and by using subtle plays on words to distract attention such as "the perfect kind of age" (Coupland, 2000, p. 12). These ads also use visual metaphor and verbal innuendo to suggest, rather than portray older adult sexual activity, such as dancing and bathing. The next section of this chapter moves away from advertisements' portrayals of older adult sexuality to that in film and television.

Seniors in Film and Television

Movie portrayals of older adults have historically conveyed negative cultural stereotypes. An analysis of the top 100 grossing films from 1940-1980 found that not only are seniors less likely to be portrayed, but also they are represented in negative ways (Bazini, McIntosh, Smith, Cook, & Harris, 1997). For both men and women, the older the character, the less attractive they were rated and the less they engaged in romantic activity. Older women in particular were negatively portrayed as unfriendly, unattractive, and unintelligent.

Similar findings were revealed in a less recent study of portrayals of older adults in television (Gerbner, Gross, Signorielli, & Morgan, 1980). However, in Gerbner, et al.'s, (1980) study, male characters' occupational power and leadership increased with age. This supports the stereotype that older men can still be "attractive" by using power and influence as proxies for physical attractiveness. Women are not so lucky. As women aged, they were portrayed as more passive and less attractive.

Few scholarly studies have been conducted to examine portrayals of older adults on network television series—especially regarding their sexual and romantic relationships. An exception is Coupland (2000) who tells of an incident on BBC's award-winning TV show, *The East Enders*, where a romance between the two main characters in their 60's culminated in marriage, after many false starts, dramas, and tragedies. A scene that demurely depicted the older characters with bare shoulders embracing in bed brought criticism from the audience on another BBC show, *Points of View*, where audience members express their praise or disapproval for broadcast shows and policies. "Complaints were expressed in words such as 'revolting,' 'horrible,' and 'unnecessary.' These complaints were driven by viewers' unwillingness to view even mild sexuality involving older people" (p. 14).

In general, the portrayals of older adults' sexual and romantic relationships in advertising, film and television reflect Woodward's (1991) assertion that the dominant response to elderliness in the West is repression. The reaction Coupland (2000) describes to the *East Ender's* white-washed sex scene supports Woodward's (1991) argument of the "unwatchability" of elderly nudes. He also argues, however, that repression of elderliness occurs in many ways. Audiences see this in the chronic underrepresentation of older adults (women in particular), the morphing of age to depict "younger" older adults, the "hush-hush" way that advertisers approach sex in older adults, and the outright invisibility of older adult sex in the Olivio/Bertolli campaign (Williams, et al., 2007).

In 2014, *The Atlantic* announced with surprise that half of all broadcast television viewers are 54 years old and older (Thompson, 2014). Given the aging of the television audience and the intense need for network television to compete with other visual entertainment media (such as cable and streaming shows), we wondered if current US prime time network television has expanded its depictions of older adults, specifically romantic and sexual relationships, to more accurately represent reality or if it continues the multifaceted approach of repressing senior's sexual romance.

THEORETICAL PERSPECTIVES

To contextualize this research, we engaged in two theoretical approaches: cultivation theory and the Communication Predicament of Aging model. The first is heavily used in media scholarship; whereas, the second is primarily used in studies on aging and communication. Collectively, they grant us access in understanding how the media present older adults' sexual relationships. We begin with cultivation theory.

Cultivation Theory

The impetus of cultivation theory (Gerbner, 1969) is based on narrative theory in that humans are story telling beings. As such, given that stories are told and retold a number of times, a moral, theme, or ideology evolves. Because we learn from a variety of sources, including media, educational institutions, governmental systems, and social networks (e.g., friends, family, community opinion leaders, etc.), the stories these sources convey are important. Thus, cultivation theory is predicated on the belief that over a period of time, viewers will adopt similar ideological views perpetuated in the media. Thus, repeated mediated messages are cultivated by audiences to the point where the portrayals become naturalized and common sense. Thus, attitudes and judgments are manufactured by the media and passed on to consumers, who, in turn, embody the same or similar beliefs (see, e.g., Bryant & Miron, 2004; Gerbner, Gross, Morgan, & Signorielli, 1980; Morgan & Shanahan, 2010).

Communication Predicament of Aging

The Communication Predicament of Aging (Ryan, Giles, Bartolucci, & Henwood, 1986) model focuses on a negative spiraling cycle of communication in intergenerational encounters. First, age related stereotypes are engaged. These stereotypes negatively impact communication behaviors through under accommodation or over accommodation of communication (such as speaking loud to a person with gray hair). When the older adult acts to correct the poor accommodation (e.g., saying "I'm not deaf"), this serves to reinforce negative aging stereotypes (e.g., old people are grumpy). The strengthened stereotype leads to constrained opportunities for communication for the older adult, lessened and poorer opportunities for communication, fewer social interactions, and, finally, a loss of self-esteem.

Combined, cultivation theory and the Communication Predicament of Aging model are ideal for this research in that they both consider the foci. That is, cultivation theory is suitable for examining what mediated portrayals of older adults' romantic and sexual relationships are communicating to audiences. It remains important to expose stereotypical or counter stereotypical portrayals to understand the messages audiences may be receiving or are encouraged to believe. Moreover, the Communication Predicament of Aging model is appropriate for examining how the (mediated) social sphere reacts to or responds to negative stereotypes of older adults' sexual relationships. When juxtaposed with cultivation theory, the social reaction perpetuated in television encourages viewers to respond ideologically similar.

Research Questions

The research questions we sought to address were twofold. First, we examined how sexual relationships are portrayed between two intimate partners, at least one of which is an older adult. This was posed with our first research question:

> RQ1: How do current, popular US prime time network television programs portray sexual relationships involving older adults?

Then, we examined how these sexual relationships are viewed by others. This was addressed with our second research question:

RQ2: How do current, popular US prime time network television programs portray the social implications of older adults' sexual encounters?

METHOD

To address the research questions two methodological approaches were employed. Specifically, we used a comparative content analysis to determine whether there was statistical significance in the portrayal and frequency of senior citizens engaging in sexual activities in American prime time network television. Then, we examined those results using critical discourse analysis to allow for a deeper, richer understanding of the portrayal. We begin with an explanation of our content analysis procedures.

Sample

To select our texts for analyses, we used a purposive sample of currently broadcast, prime time, popular American network television programs, airing from July 2009-June 2014. This represented the most recent five-year time frame of entire television seasons. To be included in the final sample, each episode displayed, discussed, or implied a sexual encounter where at least one member in the encounter was 60 years old or more. This was irrespective of whether the senior citizen was a primary, secondary, or walk-on character to allow the widest representation possible.

To identify series which would likely contain episodes with senior sex, we used online episode guides, IMDB.com, and other resources. We read the plotlines and searched for series with older characters either in the cast, as recurring characters, or guest appearances as cameos in the show. Then we reviewed episode guides to determine if the older adults were involved in romantic or sexual relationships. Initially, we identified 10 currently running comedy and drama series that met these criteria. The shows included the police drama *Blue Bloods*, the medical drama *Grey's Anatomy* and several situation comedies such as *Hot in Cleveland, Mike and Molly*, *Modern Family*, *Mom*, *The Miller's*, *The Goldberg's*, *Two and a Half Men,* and *Two Broke Girls*.

To further narrow the search, we closely examined several online episode guides to identify specific episodes from these series to code which seemed to portray a senior character in a sexual relationship. We identified 55 episodes to watch. Several episodes (n = 15), and thus four series (*Mom*, *The Goldberg's*, *The Miller's*, and *Two Broke Girls*) were eliminated from the sample. While these episodes showed older adult characters in romantic or potentially romantic situations (going on a date, being set up for a date), they did not discuss, display, nor imply sexual behavior involving seniors.

The final sample consisted of 40 episodes from 6 series[2]. Some of the shows under analysis had a narrative arc with reoccurring characters' relationships, other shows included episodes with a single romantic interlude, whether with a reoccurring character or a cameo appearance, and some shows had sexual encounters with more than one senior couple. The unit of analysis for this study was the romantic couple in each episode of each television program. Accordingly, some episodes were coded once for one couple and again for another couple.

Coding and Intercoder Reliability

A coding sheet and code book were developed to code the following: Series, Season/Episode, Genre, Character Name, Characters' Age, Sex, Marital Status, Jealousy, Display of Sexual Intimacy (*shown, discussed*), Sexual Performance, Relational Commitment, Monogamy/Promiscuity, Communication Style (*assertive, aggressive, passive/aggressive, passive*), Relational Social Integration (*public vs. private*), Social Consequences of Revealing Relationship, and Reasons for Keeping Relationship Private (if applicable) (see Appendix 1 for Code Book). All episodes were coded independently by both authors. Differences were resolved by discussion and re-watching the episodes in question.

Cohen's kappa of .80 was set as the minimum acceptable intercoder reliability for this study. Overall reliability across categories (k = .97) as well as for individual categories exceeded this minimum criterion. All categories showed an intercoder reliability greater than k = .90 except Relational Social Integration (k = .82), and Social Consequences (k = .87), which were still above the acceptable minimum.

Critical Discourse Analysis

The results of the comparative content analysis were further explored with an ideological critical discourse analysis (see Fairclough, 1992; Fowler, 1991; van Dijk, 1993, 1996). Pointedly, the content analysis findings were critically analyzed to deeper understand the portrayal of older adults' sexual relationships. To do this we drew upon Fiske's (1987) notion of discourse. As he explains, discourse is a topic or coded set of messages about a topic. For the purposes of this research, the discourse, or topic, we examined was older adults' sexual relationships.

Thus, to critically interrogate the discourse of older adults' sexual relationships in popular prime time American television, we took copious notes on the scenes that enacted this discourse. This included the characters' verbal and nonverbal cues such as dialogue, actions, reactions, body language, facial expressions, and attire. Moreover, we also noted production elements such as laugh tracks, camera angle and depth of field to illuminate any clues. Combined, these cues comprised an ideological perspective that, according to Gerbner's (1969) cultivation theory, becomes naturalized over time. Naturalization is the process in which repeated messages become accepted as normal and conventional. Hence, using this approach allowed us to determine how society is reflected in television and, simultaneously, how television shapes and affirms existing social structures (Teo, 2000). Thus, in examining the discourse of older adults' sexual relationships, we uncovered ideological assumptions and conceived of how the discourse is reified and perpetuated in popular American prime time network television.

Results

According to our study, seniors' sex is rarely portrayed on network television series. This bolsters Bell's (1992) research examining the five most watched by the elderly prime-time television programs in the United States. Of the 647 total episodes contained within the 6 series we examined which had older characters engaging in sexual relationships, only 0.06% (n = 40) episodes portrayed older adults either having sex or discussing their own sexuality. This number is shockingly low compared to the number of seniors who are sexually active discussed above (see Lee, et al., 2015; Lindau, et al., 2007) where seniors report having sex at least two or three times per month.

This number is also very low compared to the portrayal of sex on television in general. In several analyses of US prime time television by Kunkel and his colleagues, sex talk occurred in 57%-67% of the programs (Kunkel, Cope, & Biely, 1999; Kunkel, Eyal, Donnerstein, Farrar, Biely, & Rideout, 2007). Sexual behaviors occurred in 23% of the shows in 1998 (Kunkel, Cope, Farinola, Biely, Rollin, & Donnerstein, 1999) and nearly doubled in frequency by 2005 (Kunkel, Eyal, Finnerty, Biely, & Donnerstein, 2005). Other content analyses of US television found increasingly frequent reports of sexual talk and behaviors. For example, Fisher, Hill, Grube, and Gruber (2004) discovered that 82% of the scripted programming examined contained sexual talk or behavior.

The amount of senior sex portrayed is also quite low compared to sexual portrayals on television outside of the US. An analysis of British soap operas found that sexual talk or sexual intimacy occurred in 6.5 scenes per hour (Al-Sayed & Gunter, 2012). Eyal, Raz, and Levi (2014) found that over 80% of programs on Israeli television had at least one scene with sexual content.

Our first research question sought to understand how current, popular US prime time network television programs portray sexual relationships involving older adults. We found that despite the paucity of senior sex illustrated on popular US network television, it occurred primarily among heterosexual participants close to the same age, who were divorced or widowed (see Table 1), and complimented each other's sexual performance (see Table 2). Moreover, the sexual seniors were always of the same race, they did not display jealousy, but portrayed a range of relational commitment.

To address our second research question which inquired how these sexual relationships were viewed by others, in terms of their relational social integration, even though their relationships were predominantly public, and most other characters were supportive of their sexual activity, 41.8% of the time seniors' family members were shown either being unsupportive or having mixed feelings about the sexual relationship. Hence, the social implications of older adults' sexual encounters suggest that in US culture, a large number of people do not want to acknowledge senior sex. Details and discussions of these findings comprise the remainder of this chapter.

Seniors' Sexual Orientation

A total of 43 senior sexual encounters were identified among 33 characters (n = 16 males, n = 17 females) in the 40 episodes. Most sexual encounters were heterosexual (86.0%). Of the 5 characters who engaged in same-sex relationships, all of them were women, and all appeared in 4 episodes of the comedy *Two and a Half Men*. An analysis of type of sexual encounter by genre was not significant ($Chi^2 = 3.37$, $p = .07$, see Table 3). However, 14% of the senior sexual relationships overall and 26.1% of the senior sexual relationships on comedies were lesbian encounters. These depictions show a higher incidence of same-sex relationships than is reflected in the population. According to a recent analysis of several demographic surveys by the Williams Institute (Gates, 2014), only between 2.2-4.0% of adults identify as LGB/T. The depiction of senior same-sex relationships is also somewhat higher than the depiction rate of gay and lesbian relationships found on television in general. Fisher, Hill, Grube, and Gruber's (2007) analysis of television programming found that about 15% of TV series portrayed GLB sex compared to 16.7% of the TV series in this investigation.

Table 1. Categories, Frequencies and Percentages

	n	%		n	%
Episodes by Series (N = 40)			Sexual Intimacy Portrayal (N = 43)	20	46.5
Two and a Half Men	11	27.5	Shown	14	32.6
Grey's Anatomy	10	25.0	Discussed (not shown)	9	20.9
Hot in Cleveland	9	22.5	Neither		
Mike and Molly	5	12.5	A's Commitment Level to B (N = 43)		
Blue Bloods	4	10.0	1 (Low)	5	11.6
Modern Family	1	2.5	2	3	7.0
Episode by Genre (N = 40)			3	8	18.6
Comedy	26	65.0	4	3	7.0
Drama	14	35.0	5 (High)	21	48.8
Character Sex (N = 33)			Unknown	3	7.0
Female	17	51.5	B's Commitment Level to A (N = 43)		
Male	16	48.5	1 (Low)	6	14.0
Character Age (N = 33)*			2	1	2.3
20's	2	6.1	3	8	18.6
30's	4	12.1	4	0	0

Table 1. (Continued)

	n	%		n	%
40's	2	6.1	5 (High)	25	58.1
50's	2	6.1	Unknown	3	7.0
60's	16	48.5	Relational social integration (N = 43)		
70's	2	6.1	Public	37	86.0
80's+	5	15.1	Private	5	11.6
Marital Status of Characters (N = 86)†			Both	1	2.4
Single (never married)	4	4.7	Relationship revealed (N = 43)		
Married	10	11.6	Intentionally	34	70.1
Divorced	27	31.4	Accidentally	5	11.6
Separated	2	2.3	Unknown	4	9.3
Widowed	37	43.0	Social consequences of the relationship (N = 43)		
Unknown	6	7.0	Supportive	21	48.8
Type of Sexual Encounter (N = 43)			Unsupportive	11	25.6
Heterosexual	37	86.0	Mixture	7	16.3
Lesbian	6	14.0	Unknown	4	9.3

* Two characters aged over the course of the seasons and went from being considered in their 60's to their 70's. In this table, one character is represented as being in her 60's while the other is represented as being in her 70's. † Marital status is reported by episode because several characters' marital status changed as more information was learned about them. For example, one character changed from being widowed, to married, to separated, to divorced in a sequence of episodes.

Table 2. Sexual Performance of Character A by That of Character B

	Sexual Performance of Character A			Total
Sexual Performance of Character B	Stamina (3)	Neither (2)	Impotence (1)	
Stamina (3)	17	0	1	18
Neither (2)	0	24	0	24
Impotence (1)	1	0	0	1
Total	18	24	1	42*

Spearman Rho = 0.776, df = 40, $p \leq .0001$. * One couple's sexual performance was unable to be determined.

Table 3. Type of Sexual Encounter by Genre

	Genre		Total
	Drama	Comedy	
Type of Sexual Encounter			
Heterosexual	14	23	37
Homosexual	0	6	6
Total	14	29	43

$\chi^2 = 3.3663$, $P = .067$.

Evelyn Harper is seen in two of the three lesbian relationship episodes of *Two and a Half Men*. She is a character who seemingly beholds questionable scruples as evidenced from her sons (Season 8, Episode 8; Season 9, Episode 19; Season 9, Episode 20). In Season 8, for instance, her sons, Charlie and Alan, attempted to set her up with Russell, the pharmacist, only to find out he was in a relationship with a woman about half his age, Jill. Towards the end of the evening, Charlie and Alan went looking for Evelyn. Through the bathroom door, they heard the shower running and Evelyn giggled, "Oh that tickles." To which the young woman responded, "You started it. Happy birthday." Evelyn ached, "Oh Jill." Jill responded with the sound of joyous surprise, "Oh Evelyn!" The women laughed some more while viewers saw Alan and Charlie run out of the room. Moments later, Alan unsupportively commented, "I can't believe mom ran off with this idiot's girlfriend."

In the next season, Evelyn engaged in a sexual relationship with Jean, the mother of her son's (Alan) girlfriend (Lindsey). Even though both women had been married to men and were widowed or divorced, they met while at dinner with their children and apparently enjoyed each other's company. Both women talked about their attraction to men. For instance, Evelyn disclosed that one of the things she likes about travelling is the "uncircumcised men." More tamely Jean asked, "Do you think we'll find some cute fellows to dance with?" The next morning at Alan's girlfriend's house, Jean entered the kitchen in her bathrobe and informed Alan and Lindsey, "I adore your mother…I think I got to know her pretty well." At that time, Evelyn entered the kitchen in her bathrobe and affectionately said, "Good morning darling." To which Jean responded, "Hello Princess." They embraced and kissed. Alan turned to a stunned Lindsey and said, "Once again, I am so, so sorry." Their relationship continued in the next episode as well.

Two seasons later, Evelyn's granddaughter, Jenny Harper became a regular on the show and was openly gay. While her sexual exploits were akin to her late father's insatiable sexual conquests, she engaged in a three-way sexual relationship with two senior women in one episode (Season 11, Episode 6). Specifically, Alan learned that his housemate Walden Schmidt knew actress Linda Carter and wanted to be introduced to her. Walden finally gave in and invited Linda and his mother, Robin Schmidt over to the house. While Alan worked hard to impress her, his niece, Jenny, also made sexual advances on her. Linda explained to Walden, "It seems that everyone in this house wants to be with me except the one that I want." Walden laughed and then asked, "Wait. What?" At that time, Linda wrapped her arms around him and kissed him passionately. The next morning, Walden and Alan talked on the balcony while in the background viewers saw Jenny in a nightshirt kissing Robin, who was wearing the same clothes from the prior night, and quietly sneaking her out the front door. Then, Linda entered also wearing the same clothes from the night before and kissed Jenny as she quietly departed the house.

In all these instances, the seniors who engaged in lesbian sex (Evelyn Harper, Jean, Linda Carter, and Robin Schmidt) were apparently bisexual in that they also displayed sexual interest with men as well. In fact, three of them were married to men and had children. Linda Carter's status was not discussed during the show but since she is a real celebrity, viewers may know that she is married to a man and has two children. Notably, she is also a gay rights activist.

Intergenerational Integration

As briefly evidenced from the examples above, a range of character ages were portrayed. Just in *Two and a Half Men* alone, Russell, the 60-year-old pharmacist was dating Jill, a 30-year-old out-of-work actress and Jenny, the 30-year-old niece, made sexual advances toward Linda Carter, a 60-year-old actress. Our data revealed characters who engaged in sexual behavior with seniors were in their 20's (n = 2), 30's (n = 4), 40's (n = 2), 50's (n = 2), 60's (n = 16), 70's (n = 2), and 80's and older (n = 5). Of the 33 separate characters involved in senior sexual relationships, 23 characters were 60 years and older. Two characters moved from being considered in their 60's to their 70's in later seasons of the show. Both of these characters were portrayed as lovers in *Two and a Half Men*: Evelyn Harper (played by Holland Taylor) and Jean (played by Georgia Engle).

Even though a range of ages were illustrated, most sexual partners were considered to be about the same age. The vast majority of couples (72%) were rated to be in the same age category (58.1%) or no more than 1 age group apart (14%). All of the larger age differences (n = 10) were found in comedies; 8 of them were seen in *Two and a Half Men*, 1 in *Hot in Cleveland*, and 1 in *Modern Family*. This means that comedies portrayed large age differences 41.4% of the time.

For the sexual relationships involving characters with a larger gap in age, 4 of them showed older women with younger men (*Hot in Cleveland* and *Two and a Half Men*). This plays off of the "cougar" stereotype where women "prey upon men almost young enough to be their sons. Some even go after their daughters' boyfriends" (Lowen, 2015, p. 1). In fact, in *Two and a Half Men*, Walden had sex with Alan's mom (Season 9, Episode 7) and Alan had sex with Walden's mom (Season 9, Episode 13). In the other instance, Walden had sex with his date's mom (Season 10, Episode 23).

Older men were depicted in sexual activity with younger women in 3 instances (*Modern Family* and *Two and a Half Men*). This is part of an ongoing storyline for *Modern Family* as Jay negotiates his life and family with his younger wife, Gloria. However, in *Two and a Half Men*, the characters were walk-on's and not seen again in the series. For instance, Charlie and Alan wanted to set up their mom, Evelyn, with Russell (in his 60s) only to find out he was dating the much younger Jill (Season 8, Episode 8). In another episode Alan had sex with Megan (in her 30s) and learned that she was married to a much older man, Victor (Season 10, Episode 22). Rather than highlighting the "dirty old man" stereotype, the episodes focused on the young women portrayed as gold diggers. For instance, Jill explained, "We have an arrangement. I let him climb on top of me a couple of times a month and he pays my rent" (Season 8, Episode 8). Megan disclosed: "You'll have to understand; Victor was 80 when I married him. And now, twelve years later he's still alive." She met him while pole dancing in a strip club and continued, "I can't expect you to understand what it's like to rely on someone to put a roof over your head and provide you with a luxurious life you deserve" (Season 10, Episode 22). Even though Russell bragged to Charlie and Alan, "I can't believe I'm hitting that" as he pointed to Jill leaving the room, Victor was comatose in a wheelchair throughout most of the episode. Hence, the focus was not on the men's sexual conquests but rather on the younger women's incentives to have relationships with them.

The remaining three instances of seniors engaging in sex with much younger people were lesbian encounters in *Two and a Half Men*. Specifically,

Evelyn (in her 60s) engaged in shower sex with Jill by the end of the episode (Season 8, Episode 8) and Jenny (in her 30s) had an implied three-way sex with Robin Schmidt and Linda Carter (both in their 60s). Absent was any discussion regarding their incentives to do so (e.g., money, ego, etc.) but rather seemingly engaged in lesbian sex for sheer pleasure and desire.

Marital Status

Similar to age ranges portrayed, a variety of marital statuses were displayed as well. The episodes showed those engaged in senior sex were portrayed as single (never married) (n = 4), married (n = 10), divorced (n = 27), separated (n = 2), widowed (n = 37), and unknown (n = 6) (See Table 1). Four characters displayed various martial statuses as the shows progressed. Evelyn Harper (*Two and a Half Men*) was shown as both divorced and widowed during the series. Richard Webber (*Greys Anatomy*) went from being married, to separated, then widowed. His relational partner, Catherine Avery, was divorced at one time and widowed at another. In *Hot in Cleveland*, Elka Ostrovsky's marital status was used as a comedic plot device. For decades she thought her husband, Bobby, dead until he showed up at and stopped her wedding to Max. Later, after trying to make their marriage work, Elka and Bobby divorced.

The fact that a majority of the characters in our study were divorced or widowed coincides with Bidtgård's (2000) examination of nine Swedish films that cast elderly people as leading characters. In the study, Bidtgård focused on elderly people's sexuality and identified five "rules of representation" that transcended the films. Rule 5, the author explains is "sexually active elderly people have to be single" (p. 176), typically as widows or widowers. Our research slightly departs from this since some of the characters under investigation were divorced.

Whereas the sexually active seniors in our research were divorced or widowed, there were two instances in which the couples were in long-term relationships (*Grey's Anatomy* and *Modern Family*). As Bidtgård (2000) elucidates, "sex in the context of a long-term relationship is totally absent" (p. 176), and our findings are comparable. In fact, our findings align with Bidtgård's findings in that the couples' sex lives had apparently gone stagnant and/or they were seeking ways to make it exciting. In *Grey's Anatomy*, for instance, Mrs. Olsen was trying to spice up their sex life by taking pole dancing lessons to rouse her husband. Unfortunately for both of them, while attempting the "sexy flexy" she fell on her husband and they both ended up in the hospital, her with a bruised tailbone and he with possible internal injuries. In *Modern Family*'s Valentine's Day episode (Season 4, Episode 15), Jay and Gloria's sexual interlude continually got interrupted by their 6-week-old son, their extended family, and a repairman. By the end of the episode they sat in bed hearing about her son's Valentine's Day dance rather than sharing an intimate sexual experience. Thus, while divorced and widowed seniors actively engaged in sexual activity, the seniors in long-term relationships did not fare as well.

Sexual Compatibility

In terms of sexual performance, partners were almost always shown as matched (*Spearman Rho* = 0.776, df = 40, $p \leq .0001$, see Table 2). Only 2 cases showed one partner as sexually satisfied while the other was not. Most of the sexual performances (55.8%) were coded neither as impotent or possessing stamina but rather ordinary, typical, and average. However, 39.5% of the time, the sex acts were portrayed as displaying "stamina" with multiple sex acts, vigorous sex, innovative sex, or especially satisfying sex.

Our findings of sexual performance go against cultural stereotypes that older adults are not sexually capable. Despite what advertisers of Viagra and Cialis would have us believe, the seniors in our study were not only sexually capable, 42.9% of the senior sexual portrayals were displayed stamina by both partners. Only one episode even mentioned requiring "pills" for sexual performance (*Two and a Half Men*, Season 8, Episode 8).

Sexual performance differed by genre. Not surprising, only 2 dramatic episodes featured sex with stamina, whereas, 16 instances were identified in sitcoms. Of the two in dramatic series (*Grey's Anatomy*), both partners displayed it in one episode (Drs. Webber and Avery, Season 9, Episode 4), and one woman evidenced it in another episode (Mrs. Olsen, Season 9, Episode 11). All the sitcoms under investigation had at least one occurrence of a character's sexual stamina. Yet, while in 15 of the instances both partners evinced stamina, in 1 case a woman revealed stamina and her male partner did not (Joyce in *Mike and Molly*, Season 2, Episode 14). That women were shown possessing sexual stamina supports the notion that older men are not as interested or adventurous as their counterpart. This may be due to lower levels of testosterone or inability to perform.

One example of sexual stamina was exhibited by Elka and her long-estranged husband, Bobby, in *Hot in Cleveland* (Season 3, Episode 1). While on the couch, she explained that she does not want to be with him. As they began a disagreement, Bobby said "It's happening. Isn't it? The heat." To which Elka responded, "I'm on fire." Bobby retorted, "I could roast marshmallows over here." Elka asked, "What are we going to do about it?" Bobby, "Everything." Elka conceded, "Damn you, bald, ancient, Adonis!" Their tense body language, facial expressions, and the canned laugh tracks combined suggest that they possessed a mutually compatible voracious sex life. Later, Elka disclosed, "Other than the raging passion, we don't have a thing in common…All we had was 'the heat.' I will miss that."

Another example of compatible stamina existed with Joyce and Vince in *Mike and Molly* (Season 2, Episode 14) when he explained to Mike that Joyce "is a sexual volcano. Insatiable." Vince continued, "One time we were at the Olive Garden and she took her shoe off under the table. And I don't know how she does it, but her big toe and second toe form almost a perfect circle. If I didn't know any better, I'd swear it was my own hand." Noticeably US censorship does not allow prime time network television to show explicit sexual activity; however, it provides ample fodder for situation comedy storylines. This is particularly comedic when juxtaposed with older adults since stereotypes suggest that they are not interested or able to sexually perform.

Senior Sex and Social Acceptance

Not only is it important to understand how the discourse of senior sex is portrayed in television, but it is also imperative to comprehend the social implications as well. That is, we coded whether the relationships were public or private and if they were public, we focused on other characters' reactions to the older adult sexual relationships. Of the older adult sexual encounters examined, we found that 86% (n = 37) of them were public relationships.

The relationships that were kept quiet (n = 5), all appeared in dramas (*Blue Bloods* and *Grey's Anatomy*). In only three instances, reasons for their discretion were made clear to viewers. In two episodes of *Blue Bloods*, Police Commissioner Frank Reagan (played by Tom Selleck) and reporter Kelly Davidson (played by Andrea Roth) revealed that their personal relationship maintained a conflict of interest for their professional relationship (Season 1, Episode 2; Season 1, Episode 3). Ultimately, however, their relationship did

not endure. In the second season of the same series, Commissioner Reagan similarly developed a relationship with television reporter, Melony Mains and viewers learned that he "preferred to keep their friendship quiet" (Season 2, Episode 16), again suggesting that he was aware of the potential conflict of interest.

Likewise, in *Grey's Anatomy*, initially the romantic and sexual relationship between Chief of Surgery Richard Webber and Dr. Catherine Avery was kept private; however, viewers were not explicitly told why. Initially, Richard's estranged wife was diagnosed with Alzheimer's and was eventually placed in an extended care home. While romantically and sexually interested in Catherine, he was fraught with guilt since he was technically still married. After his wife passed away, he pursued a relationship with Catherine but they continued doing so in private, presumably because Catherine's son worked as an intern under Richard. Eventually, their relationship became public (see Season 8, Episodes 21, 22; Season 9, Episodes 4, 11, 16, 17; Season 10, Episodes 5, 16, 21).

Of the known romantic and sexual relationships among older adults in our study, the majority of them were supported (n = 21), some were unsupported (n = 11), and others had mixed reactions among characters (n = 7). Not surprising was that in most of the cases, the older adults' friends were supportive whereas the older adults' children were not. For instance, Jackson, Catherine's son in *Grey's Anatomy*, was not pleased that his mom was dating Richard (Season 8, Episode 21; Season 9, Episode 4); Danny, Frank's son in *Blue Bloods*, did not embrace his dad dating Melanie (Season 2, Episode 16). Additionally, in *Two and a Half Men*, Walden did not approve of his mother having sex with Alan (Season 9, Episode 13) and Alan did not appreciate his mom having sex with Walden (Season 9, Episode 7); Alan and his girlfriend Lindsey did not endorse their mothers' romantic and sexual relationship with each other (Season 9, Episodes 19, 20); Alan and Charlie were not happy their mom had shower sex with Jill (Season 8, Episode 8). In *Mike and Molly*, Mike was not happy to have his boss, Patrick Murphy, date his mother and engage in romantic and sexual activity with her (Season 1, Episode 23; Season 3, Episode 5). *Hot in Cleveland* was the only show that illustrated unsupportive characters who were not related to the older adult. Specifically, Elka Ostrovsky (played by Betty White) often displayed her sexual desires to her female roommates. For the most part they were happy and even sometimes jealous of Elka's sex life; however, they were not supportive of her sexual relationship with her estranged husband, Bobby (Season 3, Episode 1) or of a much younger man, Luke (Season 5, Episode 5). In this particular episode,

Elka entered the living room and announced, “I’ve taken a lover.” To which Emmy responded, “I feel sick.” Joy resonated, “We all do, honey.” Nevertheless, we could argue that Elka’s roommates serve as surrogate daughters to her.

Discussion

To recap, according to current popular US prime time television examined in our investigation, senior sex is rare but when it does occur, it is between single heterosexuals, close to the same age, and from the same race. Additionally, the seniors complement each other’s sexual performance, do not display signs of jealousy, and maintain a range of relational commitment (from one-night stands to long-term marital relationships). Equally telling is how the social implications of senior sex are portrayed. According to our research, the senior sexual relationships are predominantly public, and while friends are typically supportive, family members are not supportive.

Our findings say something not only about US situation comedies but also about US cultural attitudes about older adults. This includes what people in the US find funny, deviant, and socially acceptable or unacceptable. Indeed, popular culture clearly serves as a window into the culture from which it is spawn.

Senior sex appears to be a raison d'être for laughs on popular US prime time television. This is evident in that of the scant depictions of senior sex, most of it appears in situation comedies and is employed as a humorous device. Furthermore, it is not just senior sex that is funny, but that which features lesbians, stamina, a lack of relational commitments (e.g., one-night stands), and a wide age gap highlighting older women as cougars and younger women as gold diggers.

In explaining humor, Palmer (1987) states that “a joke is simultaneously plausible and implausible” (p. 71). Thus, the discourse of senior sex is certainly plausible (see Nussbaum, Miller-Day, & Fisher, 2009; Lee, Nazroo, O’Connor, Blake, & Pendelton, 2015, January 27; Lindau, Schumm, Laumann, Levinson, O’Muircheartaigh, & Wait, 2007); yet the implausibility stems from the asexual senior stereotype in US culture. When implausibility dominates in the narrative, absurdity abounds creating the humorous situation (Palmer, 1987). In essence, viewers are encouraged to laugh at seniors engaging in sexual activity, senior lesbians engaging in sexual activity, seniors with sexual stamina, seniors who desire little or no relational commitments, and seniors who engage in sexual activity with others who are at least 20 years younger than themselves. Since these portrayals are deemed fundamentally implausible, they are therefore absurd and considered humorous.

Not only is senior sex both implausible and absurd, but it is found deviant in US culture as well. Deviance creates a "moment of surprise" because it is "the butt of the absurd" (Palmer, 1987, p. 90). Importantly, deviance "is always, inherently, a feature of the real social world in some form or another" (Palmer, 1987, p. 90). Hence, the discourse of senior sex is a topic in US sitcoms and employed to be funny (evidenced from the genre, punch lines, and laugh tracks) because of how it is viewed socially as a deviant topic.

While sex is often a topic of jokes (Palmer, 1987), jokes frequently "derive from role reversal" (Palmer, 1987, p. 162). Based on our examination, this role reversal is evident in two instances. First, it takes place when older adults engage in sex (as opposed to younger adults) because "the grotesque body is old" (Mellencamp, 1992, p. 279). Second, the role reversal is illustrated when older women engage in sex (rather than older men) since "as women age, they are cast for roles that decrease their romantic [and sexual] possibilities" (Gerbner, Gross, Signorielli, & Morgan, 1980, p. 42). As a discourse of humor, these portrayals promote an ageist and sexist culture.

In terms of ageism and considering the Communication Predicament of Aging model, our findings suggest that the asexual senior stereotype is perpetuated in television. Specifically, since few seniors are shown engaging in romantic sexual activity and, of the few who do, are not supported by their children, they may be less and less apt to communicate about it. Moreover, as television portrays senior sex as implausible, absurd, and deviant, it may discourage older adults to communicate their romantic and sexual desires for fear of not being taken seriously.

As for sexism, older women in our study were centrally portrayed as the butt of the sexual jokes. To which, Palmer (1987) concurs, "humour functions as part of a sexist ideology" (p. 179). For instance, viewers are encouraged to laugh at older lesbians (rather than older gay men) and older women with younger men (instead of older men with younger women) having sexual encounters. Not only do these depictions marginalize older women, but they simultaneously reify gay male relationships and the "dirty old man" stereotype. An older man having romantic and sexual relations with a (much) younger woman, for example, is encouraged and acceptable in US media and US culture (also see Galician & Merskin, 2007; Jhally, 2002; Mellencamp, 1992). Indeed, such romantic and sexual relationships are not the topics of humor rather they are serious and naturalized.

For all that it offers, television cultivates myriad perceptions including notions of love, romance, and sex, among its audiences. Even something as seemingly benign as situation comedies can present powerful messages over time since the basis of comedy is to invert narrative expectations which are derived from social formations created in the culture (Palmer, 1987). Critics must remain diligent to continually question the portrayals of underrepresented and underserved groups. "Representations in this sense becomes an interpretive act that reveals a cultural practice" (Giroux, 1994, p. 91). Since "television is a machine capitalizing on (fear of) the passage of time – as aging and death" (Mellencamp, 1992, p. 280), critics should continue examining portrayals of older adults to extract the potentially harmful messages we create and perpetuate. Once critics reveal such messages, we have an opportunity to interrogate "the material structures of power they legitimate" (Giroux, 1994, p. 90).

Limitations and Directions for Future Research

Our study has several limitations. First, the programming under investigation is limited to US prime time scripted series. Thus, it does not account for portrayals of senior sex in other types of programming, such as news, TV movies, documentaries, reality TV, or daytime programming such as talk shows and soap operas. Nor does it examine portrayals in other parts of the world. Second, the coding was conducted by the authors. Although we coded programs separately, coders blind to the study might have rated the programs differently. Finally, the study was limited to a small number of programs with repeating characters. Our results might have been different if none of the characters were repeated in the study. However, given the difficulty of finding discussion or portrayal of senior sex on television in general, we felt that if we further limited our sample to non-repeating characters, we would not be left with enough episodes to conduct a meaningful analysis. In addition, we felt that coding for each sexual encounter was more representative of the repeated exposure audiences receive.

This investigation provides several directions for future research. One direction would be to expand the exploration of depictions of senior sex to a wider variety of television programming. This could be accomplished by looking at programming in another country, or at a different time of day. Similar investigations could examine how other forms of mass media portray senior sex, such as movies, websites, advertisements, music lyrics, etc. Finally, audience effects research could study how audiences in different age groups react to discussions and portrayals of senior sex in visual media.

NOTES

1. While we acknowledge that romance and sex are two distinct activities, our focus is on sexual activity which is often a result of romance. We, primarily but not exclusively, consider the act of sex as romance actualized.
2. Even though each series is created and aired in the United States, they all have international interest. For instance, *Two and a Half Men* airs in over 50 countries, *Hot in Cleveland* as well as *Mike and Molly* airs in 7 countries, *Blue Bloods* is broadcast in 5 countries. Moreover, *Modern Family* airs in 3 countries and 3 other countries (Chile, Greece, and Iran) have bought the rights to make adaptations. Likewise, Turkey and Columbia have bought the rights to make adaptations of *Greys Anatomy* (imdb.com, 2015).

APPENDIX 1: CODE BOOK

Show Genre

a. Comedy – ½ hour scripted show based on humorous situations, usually with laugh track
b. Drama – hour scripted show about a serious subject and is not meant to make the audience laugh

For below, code only the older adult characters involved in a sexual relationship.

Older adult's character name (s). Give character name(s) of older adult in/ seeking relationship.

a. Main character name – appears regularly on show
b. Supporting character's name (or description from credits if name not given, if both are regulars, first to appear in episode)

Sex of characters A and B

- Male
- Female

Age of characters A and B: Based on information from the episode, and visual clues, estimate each character's age. a. 20's, b. 30's, c. 40's, d. 50's, e. 60's, f. 70's, g. 80+

Relational status – status of each character A and B

a. Single (never married)
b. Married
c. Divorced
d. Separated
e. Widowed
f. Unknown

Relational jealousy displayed/discussed – indicate type and character displaying jealousy (character A or B above).

a. No Jealousy displayed/discussed by either Character A or B
b. Suspicious Jealousy - by Character A or B The "suspicious" type involves primarily thoughts, behaviors, and feelings that are usually experienced in the absence of any major jealousy-evoking events. The person having high levels of anxiety, doubt, suspiciousness, personal insecurity, and also insecurity about the relationship further characterizes suspicious jealousy.
c. Reactive Jealousy- by Character A or B the "reactive" type of jealousy occurs most strongly when concrete transgressions (e.g., sexual flirting or affairs) violate critical aspects of the relationship bond between partners (e.g., expectations of sexual exclusivity). Thus, reactive jealousy is in direct response to the discovery of actual events that threaten the stability of the relationship.

Note: A prime distinction is that suspicious jealousy is related more to endogenous or internal individual factors (e.g., personal fears of insecurity or low self-esteem) whereas reactive jealousy is related more to exogenous factors that come from the social context and the relationship (e.g., actions of others in the situation or betrayal of relationship trust) (Attridge, 2013).

Sexual intimacy

a. shown (e.g., intimately touching or in situations that suggest sexual intimacy did or will take place. Not just a desire for sexual intimacy)
b. discussed (e.g., you were great last night)
c. neither (sexual intimacy neither shown, discussed or implied)

Sexual performance of characters A and B portrayed (shaking the bed), discussed ("I've been rode hard and put away wet"), or implied ("Viagra ain't what it used to be").

a. Impotence (1) – Sexual intimacy tried, but one or more partners unable to perform (lost erection, drunk, fell asleep, etc.)
b. Neither (2) – Ordinary, typical, average sex. Did not exhibit Stamina nor Impotence
c. Stamina (3) – Multiple sex acts, vigorous sex, innovative sex, especially satisfying sex

Relational commitment of A and B: based on seeing the relationship go forward long term (from Investment Model Scales' Commitment Items). 1 2 3 4 5

High commitment (5) is related to the character theoretically responding positively to these questions:

I am committed to maintaining my relationship with my partner.
I want our relationship to last for a very long time.
I feel very attached to our relationship-very strongly linked.
I want our relationship to last forever.
I am oriented toward the long-term future of my relationship.

Low commitment (1) is related to the character responding positively to these questions:

Our relationship is likely to end in the near future.
It is likely that I will date someone other than my partner.
I would not feel very upset if our relationship were to end.

Monogamy/promiscuity/celibacy: Is each character monogamous with the other or does one or another have other sexual partners (promiscuous), or is not sexual activity shown/discussed or implied for the character (celibacy)?

a. Monogamous - character has only 1 sexual partner
b. Promiscuous - character has more than 1 sexual partner
c. Celibate - no sexual activity shown/discussed or implied for the character
d. Don't Know - Not enough information to tell

Communication style of A to B and B to A

a. Assertive communication involves standing up for personal rights and expressing thoughts, feelings, and beliefs in *direct, honest, and appropriate* ways which do not violate another person's rights. Assertive communication is the preferred style ("I'm okay and you're okay."; Use of "I" statements, objective words, direct and honest statements).
    - "I am confident about who I am."
    - "I cannot control others, but I control myself."
    - "I speak clearly, honestly, and to the point."
    - "I know I have choices in my life, and I consider my options. I am fully responsible for my own happiness."
    - "We are equally entitled to express ourselves respectfully to one another."
b. Aggressive communication involves directly standing up for personal rights and expressing thoughts, feelings, and beliefs in a way which is emotionally honest, but usually inappropriate and in violation of the rights of others ("You're not okay." Views oneself as superior, Use of "you" statements, accusations, subjective, domineering, superior words).
    - The other person is inferior, wrong, and not worth anything
    - The problem is the other person's fault
    - I am superior and right
    - I will get my way regardless of the consequences
    - I am entitled, and that the other person 'owes' me
c. Passive communication involves violating your own rights by failing to express honest feelings, thoughts, and beliefs and consequently

allowing others to also violate your rights; or expressing your thoughts and feelings in such an apologetic, timid manner that others can easily disregard them ("I'm not okay, everyone else is." Use of apologetic words or at a loss for words). "I'm unable to stand up for my rights."

- "I don't know what my rights are."
- "I get stepped on by everyone."
- "I'm weak and unable to take care of myself."
- "People never consider my feelings."

d. Passive-Aggressive communication involves expressing your needs and feelings in an unclear and confusing manner ("You're not okay, but I'll let you think that I think you are." Use of sarcasm, teasing, ridiculing, false praise, and insinuations). Source: Dr. Nielson, North Dakota State University

- "I'm weak and resentful, so I sabotage, frustrate, and disrupt."
- "I'm powerless to deal with you head on so I must use Guerrilla warfare."
- "I will appear cooperative, but I'm not."

e. Don't know. The couple is not shown interacting nor discusses interaction style. Not enough verbal interaction of couple to determine.

Relational social integration (public, private)

a. Public: friends/family know about relationship and/or interact with the pair as a romantic couple.
b. Private: the couple keeps the information of their romantic relationship from friends/family.
c. Both: one character's friends/family know about their relationship (public), the other character keeps relationship private, so friends/ family don't know about the relationship.
d. Unknown. Cannot tell from the show if others outside the relationship know about it.

If Public or Both, how is the relationship made public

a. Intentionally: The character(s) voluntarily tell their social group

b. Accidentally: the relationship is unintentionally revealed to social group

If BOTH which character is Public (A or B)
Social consequences of revealing relationship (if Public or BOTH above)

a. Supportive – friends/family say and show that they approve of the relationship--and mean it ("I'm glad you're happy," "you're glowing," "it's good that you're out there")
b. Unsupportive – friends/family say/show their discomfort ("dad's too old for this sort of thing," "I don't want to picture/see that")
c. Mixture—some characters react positively (supportive) others negatively (unsupportive).
d. Don't Know – not enough information to tell

REFERENCES

Al-Sayed, R. and Gunter, B. (2012). How much sex is there in British TV? *Communications: The European Journal of Communication Research*, 37 (4), 329-344. doi:10.1515/commun-2012-0019.

Alterovitz, S. S. R., and Mendelsohn, G. A. (2013). Relationship goals of middle-aged, young-old, and old-old Internet daters: An analysis of online personal ads. *Journal of Aging Studies*, 27, 159-165. doi:10.1016/j.jaging. 2012.12.006.

Bazini, D. G., McIntosh, W. D., Smith, S. M., Cook, S. and Harris, C. (1997). The aging woman in popular film: Underrepresented, unattractive, unfriendly, and unintelligent. *Sex Roles*, 36(7/8), 531-543. doi: 10.1007/ BF02766689.

Bell, J. (1992). In search of a discourse on aging: The elderly on television. *The Gerontologist*, 32(3), 305-311.

Bidtgård, T. (2000). The sexuality of elderly people on film—Visual limitations. *Journal of Aging and Identity*, 5(3), 169-183.

Bryant, J., and Miron, D. (2004). Theory and research in mass communication. *Journal of Communication*, 54(4), 662-704. doi:10.1093/joc/54.4.662.

Coupland, J. (2000). Past the 'perfect kind of age'? Styling selves and relationships on over-50s' dating advertisements. *Journal of Communication*, 50(3), 9-30. doi:10.1111/j.1460-2466.2000.tb02850.x.

Eyal, K., Raz, Y., and Levi, M. (2014). Messages about sex on Israeli television: Comparing local and foreign programming. *Journal of Broadcasting and Electronic Media*, 58(1), 42-58. doi:10.1080/08838151.2013.875021.

Fairclough, N. L. (1992). *Discourse and social change*. Cambridge, MA: Polity Press.

Fealy, G., McNamara, M., Treacy, M. P., and Lyons, I. (2011). Constructing ageing and age identities: A case study of newspaper discourses. *Ageing and Society*, 23(3), 264-282. doi:10.1177/0261927X06289479.

Fisher, D. A., Hill, D. L., Grube, J. W., and Gruber, E. L. (2004). Sex on American television: An analysis across program genres and network types. *Journal of Broadcasting and Electronic Media*, 48(4), 529-533.

Fisher, D. A., Hill, D. L., Grube, J. W., and Gruber, E. L. (2007). Gay, lesbian, and bisexual content on television: A quantitative analysis across two seasons. *Journal of Homosexuality*, 52(3), 167-188. doi:10.1300/J082v52n03_08.

Fiske, J. (1987). *Television culture*. New York, NY: Routledge.

Fowler, K. (2006). Spending time with (a) celebrity: Sam Taylor-Wood's video portrait of David Beckham. In: S. Holmes and S. Redmond (Eds.), *Framing celebrity: New directions in celebrity culture* (pp. 241-252). New York, NY: Routledge.

Galician, M-L., and Merskin, D. L. (Eds.). (2007). *Critical thinking about sex, love, and romance in the mass media: Media literacy applications*. Mahwah, NJ: Lawrence Erlbaum.

Gates, G. J. (2014). *LGB/T demographics: Comparisons among population-based surveys*. Los Angeles, CA: Williams Institute, UCLA School of Law.

Gerbner, G. (1969). Toward 'cultural indicators': The analysis of mass mediated public message systems. *Education Technology Research and Development*, 17(2), 137-148.

Gerbner, G., Gross, L., Morgan, M., and Signorielli, N. (1980). The "mainstreaming" of America: Violence profile No. 11. *Journal of Communication*, 30(3), 10-29.

Gerbner, G., Gross, L., Signorielli, N., and Morgan, M. (1980). Aging with television: Images on television drama and conceptions of social reality. *Journal of Communication*, 30(1), 37-47.

Giroux, H. A. (1994). *Disturbing pleasures: Learning popular culture*. New York, NY: Routledge.

Imdb.com (2015). http://www.imdb.com/tv/?ref_=nv_tp_tvhm_2 Retrieved April 11, 2015.

Irvine, J. K. (2006). Selling Viagra. *Contexts*, 5(2), 39-44. doi: 10.1525/ctx. 2006.5.2.39.

Jhally, S. [Producer and Director]. (2002). *The electronic storyteller: Television and the cultivation of values.* Northampton, MA: Media Education Foundation.

Karraker, A., DeLamater, J., and Schwartz, C.R. (2011). Sexual frequency decline from midlife to later life. *The Journals of Gerontology, Series B: Psychological Sciences and Social Sciences*, 66(4), 502-512. doi:10.1093/ geronb/gbr058.

Kunkel, D., Cope, K. M., Farinola, W. J. M., Biely, E., Rollin, E., and Donerstein, E. (1999). Sex on TV: Content and context: A biennial report to the Kaiser Family Foundation. The Henry J. Kaiser Family Foundation. Available at http://kff.org/other/report/sex-on-tv-a-biennial-report-to-3/.

Kunkel, D., Cope, K. M., and Biely, E. (1999). Sexual messages on television: comparing findings from three studies. *The Journal of Sex Research*, 36 (3), 230-236.

Kunkel, D, Eyal, K., Donnerstein, E., Farrar, K. M., Biely, E., and Rideout, V. (2007). Sexual socialization messages on entertainment television: Comparing content trends 1997-2002. *Media Psychology*, 9, 595-622.

Kunkel, D., Eyal, K., Finnerty, K., Biely, E., and Donnerstein, E. (2005). Sex on TV 4: A Kaiser Family Foundation Report. Henry J. Kaiser Family Foundation. Available at https://kaiserfamilyfoundation.files.wordpress. com/2013/01/sex-on-tv-4-full-report.pdf.

Lee, D. M., Nazroo, J., O'Connor, D. B., Blake, M. and Pendelton, N. (2015, Jan 27). Sexual health and well-being among older men and women in England: Findings from the English Longitudinal Study of Ageing. *Archives of Sexual Behavior*. doi:10.1007/s10508-014-0465-1 Available at http://dx.doi.org/10.1007/s10508-014-0465-1.

Lindau, S. T., Schumm, L. P., Laumann, E. O., Levinson, W., O'Muircheartaigh, C. A., and Waite, L. J. (2007). A study of sexuality and health among older adults in the United States. *New England Journal of Medicine*, 357(8), 762-774. doi:10.1056/NEJMoa067423.

Lowen, L. (2015). What is a cougar? About.com http://womensissues.about. com/od/startingover/f/FAQCougar.htm Retrieved April 3, 2015.

McConatha, J. T., Schnell, F., and McKenna, A. (1999). Description of older adults as depicted in magazine advertisements. *Psychological Reports*, 85 (3), 1051-1056. doi:10.2466/pr0.1999.85.3.1051.

Mellencamp, P. (1992). *High anxiety: Catastrophe, scandal, age, and comedy.* Bloomington, IN: Indiana University Press.

Morgan, M., and Shanahan, J. (2010). The state of cultivation. *Journal of Broadcasting and Electronic Media*, 54, 337-355. doi:10.1080/08838151 003735018.

Nussbaum, J. F., Miller-Day, M., and Fisher, C. L. (2009). *Communication and intimacy in older adulthood.* Barcelona, Spain: Editorial Aresta.

Palmer, J. (1994). *Taking humour seriously.* London, UK: Routledge.

Palmer, J. (1987). *The logic of the absurd: On film and television comedy.* London, UK: British Film Institute.

Robinson, J. D., Skill, T., and Turner, J. W. (2004). Media usage patterns and portrayals of seniors. In: J. F. Nussbaum and J. Coupland (Eds.), *Handbook of communication and aging research* (2nd ed., pp. 423-450). Mahwah, NJ: Erlbaum.

Ryan, E. B., Giles, H., Bartolucci, G., and Henwood, K. (1986). Psycholinguistic and social psychological components of communication by and with the elderly. *Language and Communication*, 6, 1-24.

Schultz, R., Noelker, L., Rockwood, K., and Sprott, R. (Eds.). (2006). *The encyclopedia of aging* (4th ed.). New York, NY: Springer.

Teo, P. (2000). Racism in the news: A critical discourse analysis of news reporting in two Australian newspapers. *Discourse and Society*, 11(1), 7-49. doi:10.1177/0957926500011001002.

Thompson, D. (2014, March 5). Half of broadcast TV viewers are 54 and older—Yikes. *The Atlantic.* http://www.theatlantic.com/business/archive/ 2014/03/half-of-broadcast-tv-viewers-are-54-and-older-yikes/284256/.

van Dijk, T. A. (1993). *Elite discourse and racism.* Newbury Park, CA: Sage.

van Dijk, T. A. (1996). Discourse, power and access. In: R. C. Caldas-Coulthard and M. Coulthard (Eds.), *Texts and practices: Readings in critical discourse analysis* (pp. 84-104). London, UK: Routledge and Kegan Paul.

Williams, A., Ylänne, V., and Wadleigh, P. M. (2007) 'Selling the elixir of life:' Images of elderly in an *Olivio* advertising campaign. *Journal of Aging Studies*, 21(1), 1-21. doi:10.1016/j.jaging.2006.09.001.

Woodward, K. (1991). *Aging and its discontents: Freud and other fictions.* Bloomington, IN: Indiana University Press.

Zhang, Y. B., Harwood, J. Williams, A., Ylänne-McEwen, V., Wadleigh, P. M., and Thimm, C. (2006). The portrayal of older adults in advertising: A cross-national review. *Journal of Language and Social Psychology*, 25(3), 264-282. doi:10.1177/0261927X06289479.

In: Television and Romance
Editor: Amir Hetsroni

ISBN: 978-1-63485-077-3

Chapter 6

LOVE IS IN THE AIR: LOVE RELATIONSHIPS IN TELEVISION COMMERCIALS

Dorit Zimand-Sheiner and Osnat Roth-Cohen
Ariel University, Ariel, Israel

Love and marriage, love and marriage
Go together like a horse and carriage,
This I tell you brother,
You can't have one without the other.

"Love and Marriage," Songwriters:
James Van Heusen, Sammy Cahn, 1955

ABSTRACT

The creative expressions in advertisements that relate to intimate relationships and attraction between human couples, such as romance and marriage, are seldom investigated. The current research quantifies the manifestations of intimate relationships in all Israeli television commercials from 2014 (n = 2084). Only 2.5% of advertisements in the sample presented a plot about a couple's relationship. Passion (56.6%), intimacy (49%) and commitment (30.1%), which constitute the three poles of Sternberg and Weis' "triangle of love" model, were the most frequent love components depicted in advertisements that introduce a love relationship and expressions of desire were the most prominent actions depicted in these advertisements. The findings are discussed in the context of the relationships between reality and commercial media.

INTRODUCTION

Contemporary research has shed light on the presence, use, and effect of television advertisements on consumer attitudes, as well as the influence of their emotional appeals on the decision-making process. Creative appeals in advertisements such as humor, guilt, fear, and sex, are widely investigated (e.g., Eisend, 2011; Lee-Wingate, Moon & Bose, 2014; Hartmann, Apaolaza, Barrutia & Echebarria, 2014; Reichert & Lambiase, 2014). However, appeals based on intimate relationships, such as romance, have not received much research attention. Therefore, this paper attempts to quantify the manifestations of intimate relationships in television commercials in order to track the frequency of love relationship usage by advertising professionals.

WHY SHOULD WE STUDY LOVE RELATIONSHIPS IN TELEVISION ADVERTISEMENTS?

Unsurprisingly, sex is by far the most researched subject concerning love relationship in advertisements. In spite of this attention to sex in advertising, earlier studies have shown that in some western countries, only five to six percent of prime time television advertisements contain sex or nudity[1] (Maguire, Sandage & Weatherby, 2000; Hetsroni, 2009). However, other types of relationships, such as flirting or marriage, were only investigated in the context of sex-related advertising research (e.g., Pardun & Forde, 2006).

"Romance appeal" is the creative expression given in advertisements to themes of attraction between the sexes implying that buying the advertised product will improve the consumer's romantic life (Kotler, 2003). Probably for this reason, romantic love in advertising is often subsumed under sexual appeal and is not presented to researchers as an independent issue (Huang, 2004).

As far as we know, no previous content analysis has coded romantic love in television advertisements. Perhaps this is due to the difficulty in content

---

[1] These studies do not consider sex exclusively as couples' relationships, but also refer to sexy clothing, exposed body parts, nudity, etc.

analyzing love-related themes as Hetsroni demonstrated in his study of the association between television viewing and love styles (Hetsroni, 2012). This study aims to challenge the assumption of insurmountable research difficulty and determine the share of couples relationships portrayed in Israeli television advertisements, as well as the percentage of these relationships that are considered romantic love.

From a more general perspective, analyzing advertising, one of the most visible interaction sites between marketers, products and consumers, can be advantageously approached in several ways. First, since advertisements are accepted as a social tool with the potential to affect the formation of the viewer's worldview and expectations of real life relationships (Galician, 2004; Pardun & Forde, 2006), it is important to evaluate the visual and literal depictions of couples' relationships in television advertisements. Our research may help expose patterns of manipulation used to convey reality in commercial media. This kind of research helps us better understand the symbolic environment created for consumers by the media (Shoemaker & Reese, 2014). While sex and flirtation are important characteristics of romantic love (Huang, 2004), only a small fraction of the advertisements presenting sex include romance, commitment or even friendship between couples. This study, unlike earlier studies, examines different love relationship content, including friendship, sex, flirtation and romantic love, among other love relationships. This enables us to ascertain the nature of various manifestations of intimate relationships in television commercials. Since advertising can be viewed as a cultural arena, we choose to demonstrate the concept of romantic love through an examination of the love relationships as depicted in Israeli television advertisements. Television was chosen over other media, such as newspaper or radio, because of its temporal dimension which allows for a more elaborate presentation of a relationship theme. Television's advantage is also found in its integration of sound, color and visual effects that offer the viewer a multisensual experience. Finally, from the cultivation theory perspective (Gerbner, 1967, 1969), television content promotes types of love and romantic relationships which consistently affect the television viewers' perceptions and expectations about sex, love, romance and marriage (Galician, 2004). Television exposure and love styles have been widely investigated (e.g., Signorielli, 1991; Segrin & Nabi, 2002; Hetsroni, 2012), but only a few studies have dealt with the love relationship theme as presented in television content such as advertisements and series (Moore, Bensman & Van Dyke, 2006) or feature films (Johnson & Holmes, 2009).

LOVE AND ROMANTIC LOVE

In psychology and sociology, love is treated as a multifaceted experience encompassing several types of relationships. Sternberg proposed the "triangular theory of love" (1986) which includes three components: intimacy, passion, and commitment. Table 1 illustrates each of the three as one side of a triangle that describes the love between two people.

Friendship is a kind of love which is intimate. It consists of real closeness and warmth but neither passion nor commitment is present. Friendship may be equivalent to Hendrick and Hendricks' (1992) *ludus* style of love. *Infatuated love* is a love relationship composed only of the passion component, with the absence of intimacy and commitment. According to Sternberg and Weis (2006) infatuated love experiences intense emotions, which resembles Hendrick and Hendrick's conception of *eros* and the *mania* obsession. *Empty love* involves a commitment to loving someone devoid of any intimacy or passion. *Romantic love* is a love relationship that combines intimacy and passion, but lacks commitment. It shares common features with *eros* with respect to intense feelings and attraction. *Companionate love* occurs when intimacy and commitment combine into close companionship. *Fatuous love* is a love relationship with no intimacy, but it is passionate and committed. This may occur when a couple gets married quickly purely due to passion. Finally, *consummate love* is a kind of love consisting of all components: intimacy, passion, and commitment. Sternberg (1986) states that this kind of love is not too difficult to achieve, but is hard to maintain over time.

Romantic love is experienced when high levels of intimacy and passion occur simultaneously. Brehm, Miller, Perlman and Campbell (2002) argue that romantic love can be defined as the combination of liking and infatuation. An important distinction relates to commitment. Although people tend to think of commitment as one of the characteristics of romantic love, Sternberg (1986) rejects this notion. Summer love, for instance, can be very romantic even though the lovers know that it will end when the summer is over. Hence, romantic love, according to the "triangular theory of love" is characterized by high levels of intimacy and passion but a low level or even total lack of commitment. Regan, Kocan and Whitlock (1998) also include sexual attraction as one of the defining characteristics of romantic love. Misattribution and response facilitation influence passion. Romantic love also tends to impact a person's self-esteem, with the potential of increasing one's feelings of self-worth.

Table 1. The different types* of triangle of love**

Types of love relationships	Intimacy	Passion	Commitment
Friendship	Yes	No	No
Infatuated love	No	Yes	No
Empty love	No	No	Yes
Romantic love	Yes	Yes	No
Companionate love	Yes	No	Yes
Fatuous love	No	Yes	Yes
Consummate love	Yes	Yes	Yes

Note: Reprinted from *The New Psychology of Love* (p. 187) by Sternberg, R.J. and Weis, K. (2006). London: Yale University. Copyright (2006) by Yale University.

* Sternberg's theory consists of another type of relationship, i.e., "Non-love": no intimacy, no passion and no commitment. This type was not included in the analysis, since none of its components are positive and thus it actually only describes acquaintances and not a specific type of couples' relationship.

** It is noteworthy that most relationships are impure examples of the kinds mentioned above. Relationships may include all three components of the triangle, yet in different amounts.

Other studies indicate a difference between romantic love and romantic relationships. While romantic love means "the set of forces that draw members of a couple together and cause them to care about each other" (Shaver, 2006, p. 433), a romantic relationship between two adults refers to the normative pattern "when the partners… experience sexual desire, and are progressing toward marriage or another quasi-permanent status, such as cohabitation" (Berscheid, 2006, p. 417). Huang (2004) suggests that the phrase, "in a relationship" distinguishes romantic love from simple lust associated with one night stands. Hendrick and Hendrick (1992) demonstrate a connection between romantic love and lust, defining romantic love as a pleasant emotion experienced in a romantic relationship. Lazarus (1996) distinguishes between love as a sentiment, describing the love of one partner for the other, and love as an acute emotion that comes and goes in a loving relationship. Love as an acute emotion is provoked by the sight of the other, by a particular interaction, or by a romantic atmosphere. Romantic love usually, but not always, includes sexual intimacy, which Sternberg (2006) describes as passion.

With respect to marketing and value exchange, Illouz claims that modern definitions and practices of romance are intertwined with the duality of consumer capitalism (1997). Romantic love has become an intimate, indispensable part of the democratic ideal of affluence that has accompanied the emergence of the mass market, thereby offering a collective utopia cutting across and transcending social divisions. Romantic love "is a collective arena within which the social divisions and cultural contradictions of capitalism are played out" (Illouz, 1997, p. 2). Other scholars would beg to differ, and one example of a dissenting view is that of Stone (1977) who defines romantic love as the most important thing in the world, for which all other considerations, particularly material ones, should be sacrificed. Stone demonstrates the approach by which romantic love stands above the realm of commodity exchange. Yet since the 1960s, most social science disciplines have been willing to profess a correlation between culture and emotion and, by extension, between romantic love and the economy (Bourdieu, 1979; Illouz, 1997).

TELEVISION ADVERTISING AND ROMANCE

While there is disagreement over the status of love as a commodity, it is an incontrovertible fact that advertising is interconnected with promoting commodities, services, ideas, organizations, and persons (Richards & Curran, 2002). As a method of persuasive communication, advertising attempts to facilitate change among carefully identified audiences. The change may be a cognitive change in awareness and/or detailed knowledge about the advertised brand, or a change in attitude or behavior (Belch & Belch, 2009).

Due to television's unique characteristics and advantages, advertising on this medium tends to employ dramatic and stimulating lifelike depictions. This is carried out in order to attract consumer attention and to convey an outstanding experience via a combination of sight and sound (Jefkins & Yadin, 2000). Based on these factors, television advertisements may feature a dramatic narrative about a romantic relationship while minimizing the amount of information relating to the advertised product in order to entice the consumer (Kotler & Armstrong, 2008).

Reichert (2002) points to two uses of sexual appeal in advertising according to the extent it is associated with the advertiser's persuasive message contributing to the brand image. The first use is in drawing attention to the advertisement. This tactic is common with emotional appeals, when there is a low level of association between the persuasive message about the brand image and the creative appeal. For example, while the persuasive message about a beer brand might be its taste superiority, the creative appeal might be all about sexy relationships. These advertisements, characterized by low levels of association, deploy children, warm family scenes, sex or romance to gain consumer attention (Shimp, 2010). The second use of sex in advertising is highly associated with the brand image. In such cases, sex is an integral component of brand messaging. The implied message of the sexually charged advertising is that buying and/or using the brand will lead to sexually appealing outcomes (Reichert, 2002).

The use of the theme of romantic love in advertising reflects the importance of such love in the lives of consumers. According to Huang (2004, p. 53), however, an understanding the role that romantic love plays in advertising is hindered by two factors: First, romantic love is often subsumed under sexual appeal; second, not all types of romantic love are sexual in origin or purpose. Huang (2004) claims that while sexual appeal has long been studied in advertising, scant scientific reference to romantic aspects - which are not necessarily sexual, but are nonetheless present - in commercials exists.

RESEARCH QUESTIONS

This research aims to compensate for the above mentioned deficiency by investigating the manifestation of love relationships in television advertisements and the degree of romantic love in couples' relationships as featured in advertisements. Specifically, we posit four research questions (RQs):

RQ1: How frequently are love relationships featured in television advertisements?

RQ2: Which types of love relationships are depicted in television advertisements?

As Reichert (2002) suggested, there are two uses of sexual appeal in advertising according to the extent it is associated with the brand's persuasive message. Since romantic love is usually being investigated as a part of the sex appeal aspect, we suggest it has the same uses in advertisements: attracting attention (low association) and strengthen brand image (high association). Moreover, we suggest that other love relationships may have the emotional impact potential as both attention attractors and brand promise builders. Therefore:

RQ3: Is there a correlation between the various love relationships and high or low association with the brand image?

Because advertising is a reflection of the culture in which it operates, its content should be adjusted to the cultural values of the audience. Therefore:

RQ4: Is there a correlation between the source of the advertisement (local or international) and the proportion of romantic love expression in Israeli television advertisements?

METHOD

Population and Sampling

Population

This quantitative study employs a content analysis of one year's worth of television advertisements on the two main Israeli commercial channels, Channel 2 and Channel 10. We collected all the advertisements that were broadcast throughout 2014. Israel's Second Authority for Television and Radio's Digital Information Center was used as the information source (http://www.rashut2.org.il/broadcast_commercial.asp). The research corpus and the sampling method are based on Barthes' (1967) research corpus and sampling criteria: synchronicity and homogeny. According to the synchronicity criterion, all corpus material should belong to one cycle that has a natural periodic composition. One year of television advertising is considered a natural cycle. According to the homogeny criterion, all corpus material should be of the same kind. We collected all television advertisements that appeared on the two main Israeli commercial channels. The combined share of these two channels amounts to nearly half of Israeli households[2].

Sampling Procedure

After collecting all 2014 television advertisements, we identified advertisements that featured a narrative wherein a relationship between couples was portrayed. We refer here to advertisements that tell a story or

[2] The two channels together account for 49.7% of all television viewership (The Israeli Audience Research Board, 2014). According to this source, the average Israeli watches 288 minutes of television daily, which is considered relatively high figure compared to other western countries in the world. For example, the average person in the United States watched about 273 minutes of television daily in the third quarter of 2014 (time.com/3615387/tv-viewership-declining-nielsen/).

invite the audience to construct a story using the raw materials provided by the advertisement (Chang, 2009). Based on this definition, all narrative advertisements containing a plot about a couple's relationship were chosen for the analysis.

Coding

In the final stage, we coded the advertisements that portray a couple's relationship in television narrative advertisements according to Sternberg's (1986) "triangular theory of love." See Table 1 for the different types of relationships that can be generated by different combinations of components that are part of the theory. The coding book uses the triangle of love with regard to action, since "action has so many effects on a relationship," (Sternberg & Weis, 2006, p. 188). In Sternberg and Weis' view, "action" means to act in a way consistent with one's feelings. Thus, each of the three components of the triangle is depicted by a set of actions associated with feelings. By coding each component's action parameters, we are able to identify the kind of love presented in the advertisement. We are also able to distinguish romantic love, which presents intimacy and passionate actions from other love relationships including infatuated love consisting only of passionate actions that may be the equivalent of sex. All actions between couples in the advertisements were coded. The actions were coded according to the following operational definitions:

Intimacy: In a love relationship, intimacy "refers to those feelings in the relationship that promote closeness, bondedness and connectedness," (Sternberg, 1986, p. 120).

In order to identify the intimacy component in television advertising, we searched for the following actions (Sternberg & Weis, 2006, p. 187):

1. *Sharing one's possessions and time.* Example: An advertisement for insulated glass windows where a man and a woman are sitting next to each other on a living room sofa, holding hot drinks, wearing warm clothes, and smiling at each other while a child enters the room complaining that it is freezing cold.
2. *Communicating honestly with one another.* Example: An advertisement for a coffee brand where a wife tells her husband how disappointed she is in the birthday surprise he prepared for her. While

her idea of a vacation was a romantic resort, he took her to a boarding school dormitory.

3. *Expressing empathy*. Example: An advertisement for a coffee brand where a humiliated husband is called funny nicknames by his friends after his wife surprised him at his work place with a birthday cake. The wife apologizes while empathizing with his unpleasant situation.

Passion: According to Sternberg (1986), passion is the "…expression of desire and needs… [which] manifest themselves through psychological and physiological arousal" (p. 42). The coding process of passionate relationships in television advertisements included the following actions (Sternberg & Weis, 2006, p. 187):

1. *Gazing in a sexual manner*. Example: An advertisement for supplemental insurance where a man who refuses to pay extra for an insurance policy that includes a reduced price for eye glasses fails his eye exam. When he leaves the clinic, he sees an elderly woman. He asks her, "Are you on Facebook? As she turns her back to him, he gazes at her, looking up and down her body with a big smile and a nod of approval. It is understood that because he can't see clearly, he thinks she is young and attractive.
2. *Sexual touching*. Example: An advertisement for a perfume where a female model enters a photographer's studio. She has her hair and make-up completed and then stands for the shooting. The photographer approaches her, and while looking into her eyes, he moves her hair from her eyes, stroking her cheek and neckline.
3. *Making love, including sex or post-sex activity and/or the desire for sexual activity, including flirting*. Example: An advertisement for Real-Feel condoms where a cartoon female character takes a condom in her hand and approaches a cartoon character man. They start caressing and undressing each other. Every touch makes them appear less and less like cartoon characters and more like real life people.

Commitment: This component represents a decision to devote oneself to a relationship and maintain it. To code committed relationships in television advertisements, we included the following actions (Sternberg & Weis, 2006, p. 188):

1. *Marriage proposals and engagement celebrations.* Example: An advertisement for high speed internet: The setting is a big celebration. A woman with a traditional Moroccan-style engagement gown is sitting, and in front of her, her fiancé, also wearing a traditional Moroccan-style gown, is singing her a love song[3].
2. *Marriage, including the marriage ceremony and married life.* Example: An advertisement for a supermarket chain: Marital life is presented by a couple in their forties that enter a supermarket. They walk together discussing their shopping list and their chores for the week, while giving each other tips on how to save money and shop wisely.

We also coded the level of association of romance with the brand image. For instance, an advertisement for fast home delivery of heating fuel that shows a couple kissing passionately on a sofa has a low association between the passionate couple and the advertised brand image. It can be assumed that the goal of presenting the couple in this advertisement is mainly to attract viewer attention. In contrast, a commercial for Real-Feel condoms that depicts a couple that use the product and touch one another presents a high association between narrative action and the brand image because the condoms would seem to provide the couple with a heightened sense of intimacy.

Reliability

The coding process was performed by two coders who were not aware of the specific goals of the study. They were instructed by the researchers and had previously practiced coding of advertisements. The coders were trained for six hours. They worked independently and coded all of the advertisements without a specific deadline or timeframe for completion. Cohen's Kappa statistic was computed to measure agreement between coders and coding reliability (Cohen, 1960). The correlation ranged between .867 and .886 across the categories (k = .880 for advertisement source; k = .886 for intimacy; k = .861 for passion; k = .878 for commitment; k = .867 for high/low association with the brand promise). These values constitute an acceptable level of reliability in advertising content research (Kassaarjian, 1977). Cases of disagreement between coders were solved at joint discussions. In three cases

[3] The actress and actor also play a role as an engaged couple in a local TV series.

where the coders remained undecided after the discussion, the disagreements were resolved by the authors.

RESULTS

As previously mentioned, the research population consisted of all the advertisements broadcast on the two main commercial channels in Israel during 2014. In total, 2084 advertisements were collected as the research corpus. The most prominent finding was that out of those advertisements, only 53 or 2.5% of the advertisements were identified as narrative advertisements that presented a plot about a couple's relationship[4]. These findings included including romantic relationships, sexual relationships, marriage, and other relationship types. In short, the answer to RQ1 is that love relationships were seldom presented in Israeli television advertisements during 2014.

Table 2 shows the type and proportion of love relationship content in television advertisements.

From Table 2 we learn that the answer to RQ2 is that almost all love relationships which are part of Sternberg's model are presented in television advertisements, with the exception of consummate love. In four cases, none of Sternberg's three components were found, even though the advertisements were displaying a couple's relationship. Sternberg's triangle of love theory would have called it non-love, but non-love is a platonic, friendly relationship, while a close examination of the advertisements indicates that they all present a specific relationship pattern more accurately characterized as a broken relationship: either a relationship between divorced couples, or a betrayal of the relationship between married or otherwise committed couples. We suggest the addition of another category to relationships named "breaking up the commitment."

Table 2's figures clearly indicate that romantic love, amounting to 22.6% of the corpus, is second only to infatuated love that appeared in 30.1% of the television advertisements with a love relationship. While consummate love did not appear at all, empty love has the lowest percentage with a mere 3.7%, of all the sampled love relationships.

In conclusion, love relationships that include a passion component have the highest prevalence: While consummate love had null prevalence and

[4] The product type of the 53 advertisements presenting a couple's relationship content varied across 16 different categories. The variety of the categories does not allow for statistical conclusions regarding the connection between product category and other variables.

fatuous love had a lower prevalence than any other love relationships, infatuated love and romantic love had the highest prevalence across the board.

Table 2. Type and proportion of love relationship content

Type of love relationship	% of total love relationship content in ads
Friendship	11.3
Infatuated love	30.1
Empty love	3.7
Romantic love	22.7
Compassionate love	16.9
Fatuous love	7.5
Consummate love	0
Breaking up the commitment	7.5

The answer to RQ3 is that love relationships were used more with low levels of association with the brand image than with high levels of association with the brand image. Of all the advertisements that presented love relationship content, 58.5% (n = 31) presented love relationships with a low association and 41.5% (n = 22) with a high association to brand image. Nevertheless, Fisher's test with Yates' continuity correction revealed no significant correlation between low/high association and a particular love relationship.

Finally, in answer to RQ4, only the presentation of one love relationship corresponded with the advertisement's source. Fisher's test with Yates' continuity correction revealed that the percentage of infatuated love[5] in advertisements differed depending on the origin of the advertisement ($\chi^2_{(YATES)}$ = 3.71, df = 1, P = 0.05). Sixty percent of the advertisements originating from international sources contained infatuated love, while only 23.3% of Israeli-originated advertisements included infatuated love content[6].

In addition, we examined the frequency of Sternberg's model components of intimacy, passion, and commitment in television advertisements. Table 3 displays the type and proportion of the three love components.

From Table 3 we learn that passion (56.6%) is the most frequent love component depicted in the television advertisements, followed by intimacy (49%), and commitment (30.1%). In other words, expressions of desire were

[5] Which consists only of the passion component.
[6] The majority of the advertisements were local (n = 43).

the most prominent actions depicted in advertisements. On average, an advertisement contained 1.5 components.

Table 3. Type and proportion of love components content

Type of love component	% of total love component content in advertisements
Intimacy	49
Passion	56.6
Commitment	30.1

Note: The percentage of components exceeds 100% because the model predicts that more than one component might appear in every advertisement.

DISCUSSION

The most prominent finding is the low frequency (2.5%) of advertisements presenting a plot about a couple's relationship. The advertising industry in the western world commonly uses narrative in television advertisements (Stern, 1994), especially since it is evident that narrative themes positively affect the advertisement's efficiency (Escalas, 2004; Escalas & Stern, 2003). It might therefore be asked why narratives about couple's relationship are so infrequent in Israeli television advertising.

A possible explanation may be the commodification of the love components. The love characteristics - intimacy, passion, and commitment – that in the "triangle of love" model are defined and referred to a beloved person are referred to as relationship characteristics between a consumer and a desired product or a brand (Shoemaker & Reese, 2014). A strong and favorable consumer-brand relationship can generate marketing advantages such as bridging a new market, price premiums, lower marketing costs, increased market share, consumer preference or loyalty, and positive word of mouth.

By commodifying the love components, the advertiser replaces a real-life intimate relationship between human beings with a consumer-brand relationship. In fact, this symbolic environment created for consumers by the media (Shoemaker & Reese, 2014) expresses the idea of love with regard to a product or a brand and not necessarily towards a person. Brand love is defined as a reciprocal, dynamic, multiplex, and purposive relationship between satisfied consumers and their brands, and has three key components: brand intimacy, brand passion, and brand commitment (Jun, Tat, & Siqing, 2009).

This can be understood as analogous to Sternberg's (1986) love taxonomy, even though interpersonal relationships are bidirectional in contradistinction to the typical unidirectionality of consumer-object relationships. In the latter case, although obviously the object cannot love back, much less initiate a relationship, yet there are sufficient similarities to allow meaningful analogies, as firms have the capability to initiate an activity and respond to consumer behavior (Shimp & Madden, 1988).

Another explanation of the infrequent appearance of couple's relationship narratives might be that the Israeli audience is conservative and traditionally oriented. Therefore, no marketing incentive exists to publicly deal with romantic issues which should, following social dictates, be intimate and private. A high percentage of Israeli society self-identifies with the Jewish religion, although on different levels and with versatile modes of expression. Moreover, the Jewish religion plays an important role in the local Israeli legal sphere (Nehushtan, 2015). For example, Israeli state law mandates that Jewish marriage ceremonies be performed in accordance with the religious rules and restrictions set forth in Jewish law (*Halakha*). The role of religion in Israeli society might be complemented by the relatively low national divorce rate (Moshe, 2013): For example, in 2008 Israel was ranked in the 28th place in divorce rate among 38 states in the OECD.

The possibility that the working methods of local agencies play a role in the low frequency of love-relationship content is not compelling since globalization has greatly influenced the Israeli market, injecting working methods originating in international advertising conglomerates into local agencies (Roth-Cohen & Limor, 2015). Furthermore, no previous studies exist to help us determine if the Israeli advertising industry avoids using narratives in television advertisements, in general, or simply those involving couple's love relationships.

The proportion of romantic love scenarios out of all love relationships depicted in the advertisements (n = 53 in total) is 22.6% (n = 12). Although it represents the second most frequently presented content among all relationships, it is still a rare occurrence in Israeli television advertising. Advertising, as a cultural product, discourages the depiction of romantic love relationships identifiable with intimacy, passion, and commitment. Instead, a consumerist culture that seeks thrills, pleasure, and diversity, with its concomitant undermining of intimate, long-term relationships, is encouraged. This assumption needs to be validated by comparing the current research results with research in other countries where committed romantic relationships have different prevalence.

The proportion of infatuated love characterized by passionate relationships depicted in Israeli television advertising introducing love relationships is 30.1%, the highest prevalence of all love relationships presented. The passion component also enjoys the highest prevalence (50%) in the love component content in advertisements. This can be explained by the consumer-brand relationship mentioned above. Carroll and Ahuvia (2006) conceptualize brand love as the passionate emotional attachment a satisfied consumer has for the brand. This definition highlights the affective aspect of the love relationship that consists of passion as a key component. The high proportion of infatuated love can also be explained by the current advertising atmosphere which, due to increasingly crowded media space, challenges marketers with the task of breaking through the cacophony in an attempt to draw attention to their specific advertisements and brands (Dahl, Frankenberger & Manchanda, 2003). One method of drawing attention is "shockvertising," advertisements considered offensive or shocking as they violate personal and societal norms and values due to sexually inappropriate, indecent, vulgar or aesthetically unappealing content (Urwin & Venter, 2014). The higher share of infatuated compared to romantic love content may indicate a professional advertising decision to choose more sexual references, such as flirting or sexual acts, in order to attract contested consumer attention. For example, a commercial for a perfume brand that presents a man and a woman gazing at each other and kissing passionately will probably draw more attention at first glance than a commercial for a supermarket chain presenting a couple shopping together while discussing vegetable prices.

The current research findings show that love relationships are more commonly used with low association with the brands' image (58.5%). In other words, love relationships in television advertising are more commonly employed to attract the attention of audiences rather than to deliver a message about the brand's emotional or social benefits. The preference for using love relationships for the purpose of attracting consumer attention rather than transmitting real value messages might an attempt on the art of advertisers to attract the attention of consumers who try to avoid advertisements by featuring romantically appealing messages (Rumbo, 2002).

We may also suggest that the low frequency of love relationships in Israeli television advertisements can be the result of local cultural traits. Do Israelis still believe in marriage and commitment? According to earlier studies, they certainly do, or at least act as if they do. Eighty seven percent of Israeli women aged 35-39 and eighty four percent of Israeli men in the same age bracket have been married at least once (CBS, 2012). Although marriage in Israel is highly popular, the divorce rate is rising and has doubled over the past decade. The rising divorce rate might help us understand why the passion component is the most frequent love component (56.6%) presented in Israeli television advertisements. Advertising professionals probably know that passion captures their audience's heart.

Study Limitations

The research method used for the current research, like other content analysis research, is likely to shape its findings to some extent. In our case, the choice to code only narrative advertisements, and the use of coding categories derived from Sternberg's love-relationship model, which is based on actions, may have led to the omission of commercial content that may have contained love relationships in formats other than actions. For example, Campbell (2005) suggests that many advertisements present romantic scenes in the broader meaning of the term like scenes far removed from everyday life that are exotic, imaginative or idealized.

The research corpus included all the advertisements aired on the two main commercial channels in Israel throughout 2014. However, during the summer of 2014, between July 8 and August 26, a war broke out in Israel, and this fact may have affected advertisement content, causing a shift away from emotional appeals to more rational tones (Beard, 2004). In fact, all the advertisements presenting love relationships belonged to campaigns which started before or after the war. This fact may explain the relative scarcity of love relationships in television advertising during 2014.

It might be asked if the low frequency of advertisements presenting a plot about a couple's relationship constitutes a long-term trend or just a coincidence. To better understand the current study's findings, we suggest further investigation into love relationships in television advertisements, as well as a multi-national comparison. Research could then focus on the "breaking up the commitment" relationship type detected in the current study as present in 7.5% of total love relationship advertising content. This type was not included in the Sternberg's (1986) love model, yet studying the use of betrayal and the end of a relationship theme in advertising may lead us to additional insights regarding the presentation of couples' narratives in commercials.

REFERENCES

Barthes, R. (1967). *Elements of Semiology*. New York, NY: Hill and Wang, The Noonday Press (translation from French original, 1964).

Beard, F.K. (2004). Hard-Sell "killers" and Soft-Sell "Poets": Modern Advertising's Enduring Message Strategy Debate. *Journalism History*, 30 (3), 141-149.

Belch, G.E. and Belch, M.A. (2009). *Advertising and promotion*. New York: McGraw-Hill.

Berscheid, E. (2006). Seasons of the Heart, In: Mikulincer, M., and Goodman, G. S. (Eds.). *Dynamics of romantic love: Attachment, caregiving, and sex* (pp. 404-422). New York, NY: Guilford Press.

Bourdieu, P. (1979). *La Distinction: Critique sociale du jugement* (Distinction: A Social Critique of Judgment). Paris, France: Minuit.

Brehm, S.S., Miller, R. S., Perlman, D., and Campbell, S. M. (2002). *Intimate Relationships*. New York, NY: McGraw-Hill.

Carroll, B. A, Ahuvia A C (2006). Some antecedents and outcomes of brand love. *Marketing Letters*, 17, 79-89.

CBS - Central Bureau of Statistics. (2012). *Israeli Men and Women – 1990-2011* [Hebrew: Nashim u'gvarim 1990-2011]. Retrieved http://www.cbs.gov.il/www/statistical/mw2013_h.pdf.

Chang, C. (2009). "Being Hooked" By Editorial Content: The Implications for Processing Narrative Advertising. *Journal of Advertising*, 38(1), 21-34.

Cohen, J. (1960). A coefficient of agreement for nominal scales, *Educational and Psychological Measurement*, 20(1), 37-46.

Dahl, D. W., Frankenberger, K. D., and Manchanda, R. V. (2003). Does It Pay to Shock? Reactions to Shocking and Nonshocking Advertising Content among University Students. *Journal of Advertising Research*, 43, 268-280.

Eisend, M. (2011). How humor in advertising works: A meta-analytic test of alternative models. *Marketing Letters*, 22(2), 115-132.

Escalas, J.E. (2004). Imagine yourself in the product: Mental simulation, narrative transportation and persuasion. *Journal of Advertising*, 33(2), 37-48.

Escalas, J.E. and Stern, B.B. (2003). Sympathy and empathy: Emotional responses to advertising dramas. *Journal of Consumer Research*, 29, 566-578.

Galician, M.L. (2004). *Sex, love and romance in the mass media: Analysis and criticism of unrealistic portrayals and their influence*. Mahwah, NJ: Lawrence Erlbaum.

Gerbner, G. (1967). An institutional approach to mass communications research. In: L. Thayer (Ed.), *Communication theory and research: Proceedings of the first international symposium* (pp. 429-445). Springfield, IL: Charles C. Thomas.

Gerbner, G. (1969). Toward 'cultural indicators': The analysis of mass mediated public message systems. *AV Communication Review*, 17(2), 137-148.

Hartmann, P., Apaolaza, V., D'Souza, C., Barrutia, J. M., and Echebarria, C. (2014). Environmental threat appeals in green advertising: The role of fear arousal and coping efficacy. *International Journal of Advertising*, 33(4), 741-765.

Hetsroni, A. (2007). Sexual content on mainstream TV advertising: A cross-cultural comparison. *Sex roles*, 57(3-4), 201-210.

Hetsroni, A. (2009), How much sex is there really in Israeli TV commercials? A longitudinal analysis. *Social Issues in Israel*, 8, 147-174 (In Hebrew).

Hetsroni, A. (2012). The association between television viewing and love styles: A Cultivation perspective. *American Journal of Media Psychology*, 110, 35-50.

Hendrick, S. S., and Hendrick, C. (1992). *Romantic love*. Newbury Park, CA: Sage.

Huang, M. H. (2004). Romantic love and sex: Their relationship and impacts on ad attitudes. *Psychology and Marketing*, 21(1), 53-73.

Illouz, E. (1997). *Consuming the romantic utopia: Love and the cultural contradictions of capitalism*. London, England: University of California Press.

Jefkins, F. and Yadin, D. (2000). *Advertising*. London, England: Prentice Hall.

Johnson, K. R., and Holmes, B. M. (2009). Contradictory messages: A content analysis of Hollywood-produced romantic comedy feature films. *Communication Quarterly*, 57(3), 352-373.

Jun, P. A. N. G., Tat, K. H., and Siqing, P. E. N. G. (2009). Effects of advertising strategy on consumer-brand relationships: A brand love perspective. *Frontiers of Business Research in China*, 3(4), 599-620.

Kotler, P. (2003). *Marketing Management*. London, UK: Prentice Hall.

Kotler, P., and Armstrong, G. (2008). *Principle of Marketing*. London, UK: Prentice Hall.

Lazarus, R. S. (1996). *Passion and reason: Making sense of our emotions*. Oxford University Press.

Lee-Wingate, S. N., Moon, J. Y., and Bose, M. (2014). The influence of mortality focus on guilt advertising effectiveness. *Journal of Marketing Theory and Practice*, 22(1), 103-114.

Maguire, B., Sandage, D., and Weatherby, G.A. (2000). Violence, morality and television commercials. *Sociological Spectrum*, 20, 121-143.

Moore, B., Bensman, M. R, and Van Dyke, J. (2006). *Prime time television: A concise history.* Westport, CT: Praeger.

Moshe, N. (2013). A selection of data on divorce in Israel. [Hebrew: Mivchar netunim al gerushim beIsrael]. *Hakneset, center of research and information.* Retrieved http://www.knesset.gov.il/mmm/data/pdf/m03199.pdf.

Nehushtan, Y. (2015). *Intolerant religion in a tolerant-liberal democracy.* Oxford: Hart Publishing.

Pardun, C.J., and Forde, K. R. (2006). Sexual content of television commercials watched by early adolescents. In: T. Reichert and J. Lambiase (Eds.), *Sex in consumer culture: The erotic content of media and marketing appeal* (pp. 125-139). Mahwah, NJ: Lawrence Erlbaum.

Reichert, T. (2002). Sex in advertising research: A review of content, effects, and functions of sexual information in consumer advertising. *Annual review of sex research*, 13(1), 241-273.

Regan, P.C, Kocan, E.R, and Whitlock, T. (1998). Ain't love grand! A prototype analysis of the concept of romantic love. *Journal of Social and Personal Relationships*, 15, 411-420.

Richards, J.I. and Curran, C.M. (2002). Oracles on "Advertising": Searching for a Definition. *Journal of Advertising*, 31(2), 63-77.

Roth-Cohen, O. and Limor, Y. (2015). From an Art to an Industry: Milestones in the Evolution of the Israeli Advertising Industry [Hebrew: Meomanut l'etaasia: gormim, tahalichim venekudot tzion behitpatchut ta'asiat hapirsum bisrael]. *Media Frames*, 14, 58-86.

Rumbo, J. D. (2002). Consumer resistance in a world of advertising clutter: The case of Adbusters. *Psychology and Marketing*, 19(2), 127-148.

Segrin, C., and Nabi, R. L. (2002). Does television viewing cultivate unrealistic expectations about marriage? *Journal of Communication*, 52 (2), 247-263.

Shaver, R. Phillip (2006). Dynamics of romantic love. Comments, questions, and future directions. In: Mikulincer, M., and Goodman, G. S. (Eds.). *Dynamics of romantic love: Attachment, caregiving, and sex* (pp. 423-456). New York, NY: Guilford Press.

Shimp, T. A, Madden, T. J. (1988). Consumer-object relations: A conceptual framework based analogously on Sternberg's triangular theory of love. *Advances in Consumer Research*, 15, 163-168.

Shimp, T. A. (2010). *Integrated marketing communications in advertising and promotion*. Mason, OH: Thomson, South-western.

Shoemaker, P. J., and Reese, S. D. (2014). *Mediating the message in the 21st century: A media sociology perspective*. New-York, NY: Routledge.

Signorielli, N. (1991). Adolescents and Ambivalence toward Marriage: A Cultivation Analysis. *Youth and Society*, 23(1), 121-49.

Stern, B. (1991). Two pornographies: A feminist view of sex in advertising. *Advances in Consumer Research*, 18(1), 384-391.

Stern, B.B. (1994). Classical and Vignette television advertising dramas: structural models, formal analysis and consumer effects. *Journal of Consumer Research*, 20(4), 601-615.

Sternberg, R.J. (1986). A triangular theory of love. *Psychological Review*, 93, 119-135.

Sternberg, R.J. and Weis, K. (2006). *The New Psychology of Love*. London: Yale University.

Stone, L. (1977). *The Family, sex and marriage in England, 1500-1800*. New York, NY: Harper and Row.

The Israeli Audience Research Board (2014). http://www.midrug-tv.org.il/index.php?language=eng.

Urwin, B., and Venter, M. (2014). Shock Advertising: Not So Shocking Anymore. An Investigation among Generation Y. *Mediterranean Journal of Social Sciences*, 5(21), 203-214.

In: Television and Romance
Editor: Amir Hetsroni
ISBN: 978-1-63485-077-3

Chapter 7

EXPLORING THE IMPACT OF INDIVIDUAL CHARACTERISTICS AND REALITY DATING TELEVISION VIEWING MOTIVES IN PREDICTING COLLEGE STUDENTS' STEREOTYPICAL DATING ATTITUDES

Amber L. Ferris
The University of Akron Wayne College, Orrville, OH, US

ABSTRACT

The goal of this study was to determine the factors that impact reality dating show viewers' stereotypical dating attitudes. Applying a uses and gratifications theory perspective, the study explores how gender, perceived realism, romanticism, and reality dating show viewing motives relate to three dating stereotypes: men are sex-driven, women are sex-objects, and dating is a game. The results show that individual characteristics of romanticism, perceived television realism, and gender, as well as the voyeurism motive, are direct predictors of dating attitudes. Specifically, those who are less romantic are more likely to believe that dating is a game. Men, people who believe that television is realistic, and those who are motivated to view reality television for the sexual content are more likely to believe that women are sex-objects. Lastly, highly romantic people are more likely to believe that television is realistic, and people who are motivated by voyeuristic reasons to watch reality dating shows are more likely to believe that men are sex-driven. Limitations and directions for future research are also discussed.

INTRODUCTION

For over 15 years, reality dating television has been showing viewers how to date. In 1999, the first reality dating show, *Blind Date,* debuted on NBC (Huff, 2006). Differing from previous dating game shows like *The Dating Game, Blind Date* moved beyond the date selection process and filmed people as they participated in the dating process. Since the debut of *Blind Date*, there have been over 30 programs in the reality dating show (RDTV) genre, including the longest running show, *The Bachelor* (Yahr, Moore, & Chow, 2015). Now in its 19th season, *The Bachelor* premiered as the number one program for the 18-49 age demographic and averaged approximately 8 million views throughout the season (The Bachelor, 2015). New shows continue to emerge each season, including programs like VH1's *Dating Naked*, where daters are placed on an island completely naked. The season 2 finale for this program reached the top 10 most popular shows in the 18-49 demographic (Metcalf, 2015), prompting the show to be renewed for a third season (Black, 2015).

Not only are reality dating shows still going strong on traditional television, they have garnered significant social media attention. During season 19, *The Bachelor* was the number one "most chatted about" network show, overtaking the popular zombie series *The Walking Dead* (Dos Santos, 2015). After the season finale, just over 350,000 tweets were posted on Twitter regarding bachelor Chris Soules' mate choice (Flomenbaum, 2015). BRAVO TV's *The Singles Project,* a new reality dating program debuting in 2014 that follows young professional singles in New York, was specifically designed to allow viewers to react to what they are seeing and interact with cast members in real time on social media (Spangler, 2014). The integration of RDTV with social media further indicates that this genre of television is here to stay. Given the potential impact these programs might have on how we view romantic relationships and gender roles that are enacted in these relationships, continued research is warranted.

Despite being marketed as "reality," the reality dating genre often displays images and themes that are beyond what would occur on a real date, such as dating several people at the same time and taking a first date to a hot tub (Ferris, Smith, Greenberg, & Smith, 2007).

The popularity of RDTV also brought scholarly interest. So far, research has been concerned with the themes and behaviors depicted on reality dating programs, examining the impact of RDTV on beliefs about sexuality and gender roles (Rivadenrya & Lebo, 2008; Vandenbosch & Eggermont, 2011; Zurbriggen & Morgan, 2006), desirable characteristics in a potential mate (Chapdelaine, Levesque, & Cuadro, 1999; Hetsroni, 2000; Hetsroni & Bloch, 1999), and even behaviors stemming from viewing the shows such as increased likelihood of engaging in one-night stands (Fogel & Kovalenko, 2013).

One specific area of research has focused on the stereotypical portrayals of men and women, and the attitudes that relate to those portrayals. Research has shown that three stereotypical dating attitudes are related to television viewing: dating is game, men are sex-driven, and women are sex-objects (Ward, 2002; Ward & Rivadeneyra, 1999). While research has shown that increases in RDTV viewing result in greater likelihood of subscribing to these dating stereotypes (Ferris et al., 2007; Zurbriggen & Morgan, 2006), researchers must continue to explore additional important factors in the link between RDTV viewing and attitudes.

According to Papacharissi (2008), the *Uses and Gratifications* (U&G) theoretical perspective is well-suited to exploring areas where mediated and interpersonal needs can be met. In the case of reality dating programming, viewers might be drawn to the show to satisfy traditional media consumption motives like to be entertained or relieve their boredom, but also for more interpersonal needs like learning how to date. In addition to motives, U&G also proposes that individual characteristics such as demographics, personality traits, attitudes, and beliefs can impact why people seek out particular media to fulfill their needs (Rubin, 2009). The goal of this research is to apply the U&G theory to examine how viewer motivations to watch RDTV and individual characteristics are linked to viewers' stereotypical dating attitudes.

USES AND GRATIFICATIONS THEORY

The uses and gratifications theory (U&G) is a functional approach that emphasizes the individual as a central factor in media consumption (Papacharissi, 2008; Rubin, 2009). This perspective asserts that the media fulfills meaningful functions for individuals who seek out those functions based on who they are. The effects of those choices will then vary based on both the specific make-up of media consumers and what motivates them to seek out a particular channel to fulfill their needs.

Several studies have examined motives related to reality television viewing broadly, and RDTV specifically. Motives that these two genres have in common include excitement/entertainment (Godlewski & Perse, 2010; Papacharissi & Mendelson, 2007; Roberti, 2007; Vandenbosch & Eggermont, 2011; Zurbriggen & Morgan, 2008), escape (Godlewski & Perse, 2010; Roberti, 2007; Vandenbosch & Eggermont, 2011), habit/pass time (Godlewski & Perse, 2010; Papacharissi & Mendelson, 2007), and learning (Roberti, 2007; Zurbriggen & Morgan, 2008).

In addition to these common motives, some unique motives have also been reported. With respect to reality television, voyeurism (Godlewski & Perse, 2010; Papacharissi & Mendelson, 2007), companionship, social interaction, and relaxation (Papacharissi & Mendelson, 2007) have also emerged as motives. Additionally, Vandenbosch and Eggermont (2011) found the RDTV motives of mood regulation and television dependency in their research.

While several studies have focused on how motives relate to overall reality television and RDTV exposure, recent research has explored how motives can be related to attitudes and behaviors. Motives have been shown to be correlated with going online after viewing reality television programming (Godlewski & Perse, 2010), television viewing satisfaction/affinity (Godlewski & Perse, 2010; Papacharissi & Mendelson, 2007), and identification with a favorite reality TV character (Godlewski & Perse, 2010).

In their study related specifically to reality dating television, Zurbriggen and Morgan (2006) found that entertainment and learning motives were both positively related to all three stereotypical dating attitudes (dating is a game, men are sex-driven, women are sex-objects). While this study shows support for two motives related to RDTV, there is no evidence linking additional motivations for watching RDTV to adopting stereotypical dating attitudes. Therefore, the following research question is posed:

RQ1: Are RDTV motives related to the attitudes a) dating is a game, b) men are sex-driven, and c) women are sex-objects?

Individual Characteristics

According to Papacharissi (2008), individuals' demographics, psychological, and sociological characteristics influence why we choose media to fulfill our needs. The following discussion will outline three individual characteristics that are believed to impact the relationship between reality dating television motives and dating attitudes: gender, viewers' perception that television is real, and romanticism.

Gender. The impact of gender has been examined in several studies related to both television motives and dating attitudes, resulting in mixed findings. With respect to motives research, Roberti (2007) found that women were more likely to watch RDTV for excitability and escape more than men. Additionally, Zurbriggen and Morgan (2006) found that men were more likely to use RDTV to learn about dating and the opposite sex more than women. While these gender differences point to the possibility that men and women may differ in what motivates them to view reality dating television, other research has shown that there were no gender differences in learning and entertainment motives related to general television viewing (Ward, 2002).

Gender was also shown to have influence on dating and relationship attitudes. Males have been found to believe that dating is a game, that men are sex-driven, and that women are sex-objects more than females (Ferris et al., 2007; Ward, 2002). However, other research failed in detecting gender differences, both with respect to the three stereotypical dating attitudes (Zurbriggen & Morgan, 2006) and in exploring attitudes toward traditional dating gender roles (Rivadeneyra & Lebo, 2008).

Perceived TV realism. In addition to gender, perceived realism has been found to relate to both viewing motives and attitudes. Research has shown that increased perceived TV realism is positively related to the television viewing motives of entertainment, relaxation, social interaction, and companionship (Papacharissi & Mendelson, 2007; Rubin, 1981). In relation to dating attitudes, studies have shown that those who believe that reality dating shows are more realistic had more affinity for RDTV shows (Papacharissi & Mendelson, 2007), believed in less traditional dating gender roles (Rivadeneyra & Lebo, 2007), and also held stereotypical dating attitudes (Ferris et al., 2007).

Research on perceived realism and television has also shown some potential differences based on gender. Specifically, women have been reported to believe that television in general is more real (Zurbriggen & Morgan, 2006) and that dating shows represent real-life relationships more than men (Roberti, 2007). While Punyanant-Carter (2006) found similar gender results with respect to the belief that love is realistic on television, her study also found that men were more likely to believe that sexual content on TV is realistic more than women.

Romanticism. One new direction with respect to reality dating research is the inclusion of romanticism as a predictor of potential media effects. According to Sprecher and Metts (1989), romanticism is "a relatively coherent individual orientation toward love" (p. 388). While no studies have included this variable as a predictor of RDTV motives, previous research on romance and love may shed light on the potential for this variable to be impactful when examining motives and dating attitudes. For example, Hefner and Wilson (2013) found that frequent movie viewers scored higher in romanticism compared to non-frequent viewers and that those who watched movies for learning purposes were also more likely to hold stronger romantic beliefs. While that study did not find that more genre-specific movie viewing impacted romantic beliefs, some research on television genre differences has emerged. When assessing the impact of love styles and love-related television programming (including dating shows), Hetsroni (2012) found that participants who scored higher on the Eros love style (associated with romantic and passionate love) were more likely to watch romantically-themed television programming. Given that reality dating shows focus primarily on finding love and depicting dating, romanticism may also be related to RDTV viewing.

When examining the long-term impact of television viewing on adolescents' approaches to sexual initiation, Eggermont (2006) found that those who viewed more television were more likely to believe that a romantic approach to sexual requests was most appropriate. More specifically, this perception was held more by boys than by girls. These gender differences have also been found with respect to romanticism (Sprecher & Metts, 1989); specifically, men have been found to hold stronger romantic beliefs than women. These differences.

Model Testing

According to the U&G theory, individual characteristics lead to various viewing motives, which may bring about media effects (Papacharissi, 2008; Rubin, 2009) (see Figure 1). The previous research points to the potential role of gender, perceived realism, and romanticism in impacting motives and dating attitudes, but the direction of the influence is unclear. Therefore, to test the overall U&G model, the following research questions are posed:

RQ2: Which individual characteristics (gender, romanticism, perceived realism) and motives for watching reality dating television predict college students' attitude that dating is a game?

RQ3: Which individual characteristics (gender, romanticism, perceived realism) and motives for watching reality dating television predict college students' attitude that men are sex-driven?

RQ4: Which individual characteristics (gender, romanticism, perceived realism) and motives for watching reality dating television predict college students' attitude that women are sex-objects?

METHOD

Procedures

To test the study's research questions, a cross-sectional survey design was employed. Participants were recruited via CRTNET, a listserv of the National Communication Association, as well as other faculty-based social media (including LinkedIn groups). Faculty were asked to distribute the survey to students who are 18 years or older. Participants completed the survey via Qualtrics, an online survey tool. This research was approved by The University of Akron Institutional Review Board.

Participants

Completed surveys were collected from 181 students from various colleges and universities across the United States. The average age of participants was 22 years (SD = 5), with a range in age from 18 to 48 years. This age group falls directly within the target audience for many reality dating programs (see The Bachelor, 2015; Metcalf, 2015). Participants watched an average of 16.5 hours of television per week (SD = 14.6), and 90% of the sample had seen at least one reality dating show program during the past year.

The sample was predominantly Caucasian (n = 155, 85.6%) and female (n = 132, 72.9%). The sample represented various class standings, fairly evenly split between underclassmen (freshman and sophomores, n = 80, 44.2%) and upperclassmen (juniors and seniors, n = 79, 43.6%).

Other levels represented included graduate students and post-secondary high-school students attending college classes (n = 22, 12.2%).

Measures

All measures in the study achieved reliabilities consistent with previous research. Table 1 includes variable means, standard deviations, and Cronbach's alpha scale reliability scores. All scaled items were measured on a 7-point Likert-type scale, where higher numbers indicate more of an attitude or belief.

Dating Attitudes. The revised Attitudes toward Dating measure (Ferris et al., 2007) was used to assess stereotypical dating attitudes. This measure consists of three attitudes: dating is a game, men are sex-driven, and women are sex-objects. "Dating is a game" includes seven items and measures the attitude that dating is nothing to be taken seriously, and is something that often involves manipulation (α = .73).

For the "Men are sex-driven" and "Women are sex-objects" measures, participants were asked to respond to the following statements based on their thoughts related to heterosexual men and women. The "Men are sex-driven" measure consists of 10 items encompassing the attitude that men are unable to control their sexual urges and think about sex all of the time (α = .76). Lastly, the "Women are sex-objects" measures the attitude that women should focus on their looks and should expect sexual advances from men (11 items, α = .78).

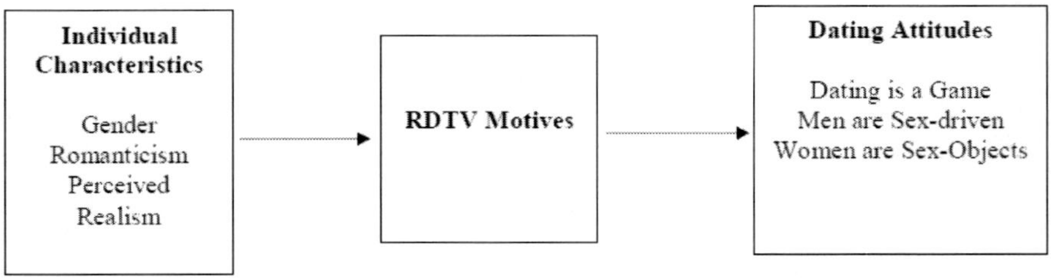

Figure 1. Proposed model of Antecedents to Dating Attitudes and Romantic Beliefs.

Table 1. Means, Standard Deviations, and Cronbach's Alphas for all Scale Measures

	M	SD	α
Dating Attitudes			
Dating is a Game	2.51	.97	.73
Men are Sex-driven	3.28	.89	.76
Women are Sex-objects	3.14	.94	.78
RDTV Motives			
Social Learning	2.32	1.28	.89
Companionship	2.17	1.28	.92
Social Interaction	3.44	1.68	.78
Voyeurism	2.86	1.65	$r = .70^{**}$
Entertainment	3.70	1.53	.92
Passing Time	4.69	1.44	.75
Romanticism	3.90	.91	.82
Perceived TV Realism	2.87	1.22	.69

Note: N = 181. The voyeurism motives were measured with two items, yielding a significant correlation.

** $p < .01$.

Reality Dating TV Viewing Motives. To measure reality dating TV viewing motives, participants were asked to indicate whether or not they had ever seen at least one reality dating show in their lifetime. If they answered no, they were taken to the end of the survey. Therefore, only those that had at least some exposure to RDTV programming completed the motives measures.

In order to measure RDTV motives, Papacharissi and Mendelson's (2007) reality television motives scale was modified to fit the context of reality dating shows. This measure included five motives: entertainment, relaxation, habitual/pass time, companionship, and voyeurism. Items were also added from Roberti's (2007) reality dating show motives scales (excitability, social learning, and escape). The full index included 30 items.

All items were subjected to an exploratory factor analysis with Varimax rotation to reveal latent factors. Eigenvalues of 1.0 or greater were necessary to be included as a factor, and the 60/40 criterion was applied to determine if items were cleanly loaded on each factor. The analysis revealed a six-factor solution including 26 final items. This solution accounted for 69.5% of the variance. See Table 2 for items, factor loadings, and variance explained by each motive. Scores were computed by summing and averaging each participant's responses on the items loaded on each factor.

Table 2. Factor Loadings for Final 6-Factor Solution of Facebook Motives

	1	2	3	4	5	6
I watch Reality Dating Shows because…						
Factor 1: Entertainment						
I just like to watch	.83					
It is exciting	.79					
It entertains me	.77					
It relaxes me	.68					
It allows me to unwind	.67					
It amuses me	.64					
It is more enjoyable than fiction programming	.63					
It is a pleasant rest	.62					
Factor 2: Companionship						
Dating shows help me forget when I am alone		.82				
It makes me feel less lonely		.82				
So I won't be alone		.80				
Dating shows give me company		.75				
Dating shows help me to forget my problems		.74				
Factor 3: Social Learning						
They help me learn how I'm supposed to act in different situations and places			.83			
They help me to learn how to do things I've never done before			.78			
They help me learn what could happen to me			.75			
They help me learn from the mistakes of others			.70			
They help me know what's going on in the world			.62			

	1	2	3	4	5	6
Factor 4: Passing Time						
Just because it's there				.81		
When I have nothing better to do				.72		
It gives me something to do to occupy my time				.67		
Factor 5: Social Interaction						
So I can be with other family or friends					.76	
So I can talk with other people about what's on TV					.69	
It is something to do when friends come over					.68	
Factor 6: Voyeurism						
I enjoy the sexual content.						.71
The program has sex appeal						.70
Variance Explained (%)	18.41	14.78	12.90	8.14	8.08	7.19

Note: N = 181. Factor loadings and variance reported from the final, rotated factor solution.

Factor 1, *Entertainment*, measures one's motivation to watch RDTV for fun and relaxation. Factor 2, *Companionship,* consists of items measuring one's motive to watch RDTV to relieve loneliness. The third factor that emerged was *Social Learning*. This factor measures the extent to which participants use RDTV to learn about themselves and others. *Passing Time* emerged as the fourth factor, measuring motive to watch dating shows to relieve boredom. Factor 5 is *Social Interaction.* This factor measures the motivation to watch RDTV in order to spend time with others or be able to talk about the programming with others. The last factor, *Voyeurism,* focuses on the motive to watch dating shows for the sexual content depicted.

Gender. Participants were asked to identify their gender as male, female, or transgender. An "other" option was also available.

Perceived Realism. Perceived realism was measured using Rubin's (1981) *Perceived Realism Scale*. This scale measures the degree to which participants believe that programming on television shows life as it really is (α = .69). The negatively worded items were removed due to the lack of reliability, thus resulting in a three-item measure: "television presents things as they really are in life," "television lets me really see how other people live," and "television lets me see what happens in other places as if I were really there."

Romanticism. Sprecher and Metts' (1989) *Romantic Beliefs Scale* was used to assess participant's romanticism. In their development of this measure, Sprecher and Metts (1989) found that the total measure of romantic beliefs, consisting of 15 items, was found to be a valid and reliable measure. Therefore, the romanticism measure was also constructed by summing and averaging all the items in the scale ($\alpha = .82$).

RESULTS

The Relationship between Motives and Dating Attitudes

RQ1 was proposed to determine how motives are related to each of the three dating attitudes: dating is a game, men are sex-driven, and women are sex-objects. To answer this question, Pearson correlations were calculated for each motive in relation to each of the three dating attitudes (see Table 3).

Overall, five of the six RDTV motives were found to be related to at least one dating attitude. For the dating is a game attitude, three motives demonstrated significant relationships: companionship, social learning, and voyeurism. Five of the six motives were related to the attitude that men are sex-driven: companionship, social learning, social interaction, voyeurism, and entertainment. Lastly, the attitude that women are sex-driven was related to three motives: companionship, social learning, and voyeurism. The passing time motive was not found to be related to RDTV motives.

U&G Model Testing

RQ2, RQ3 and RQ4 were proposed to test the U&G model with respect to how individual characteristics and motives predict each of the three dating attitudes. In order to answer these questions, path analyses were conducted. This process has been used in previous U&G research to reveal both direct and indirect predictors related to the dependent variable (Haridakis & Rubin, 2003). To conduct the path analysis, six separate regressions were run to determine the predictors of RDTV motives. In these analyses, motives were regressed on the individual characteristics of gender, romanticism, and perceived TV realism (see Table 4).

Table 3. Correlations between RDTV Motives and Dating Attitudes

	Dating is a Game	Men are Sex-driven	Women are Sex-objects
Companionship	.27**	.37**	.26**
Social Learning	.29**	.35**	.26**
Social Interaction	.05	.25**	.06
Voyeurism	.27**	.38**	.29**
Passing time	.11	.08	-.01
Entertainment	.11	.21**	.12

Note: $N = 181$. $^{**}p < .01$.

Table 4. Regressing RDTV Motives on Individual Variables

	Voyeurism	Companionship	Social Learning	Social Interaction	Passing Time	Entertainment
Predictors	Final β	Final β	Final β	Final β	Final β	Final β
Gender	-.07	.08	-.03	.19*	.27***	.32***
Perceived Realism	.22**	.43***	.47***	.18*	.10	.25***
Romanticism	.01	.07	.07	.13	.12	.10
Model R^2	.06*	.20***	.24***	.08**	.09**	.16***

Note: $N = 181$. All betas are standardized betas.
$^{*}p < .05$. $^{**}p < .01$. $^{***}p < .001$.

Next, hierarchical regressions were run on each of the three stereotypical dating attitudes. Variables were entered into each step in accordance with U&G theory (see Figure 1). Block one included gender, romanticism, and perceived realism. Block two included each of the six RDTV motives (see Table 5).

Dating is a Game Attitude. RQ2 was proposed to determine the predictors associated with the dating is a game attitude. The total model explained 23% of the variance related to participants' dating is a game attitude, $F_{(9, 169)} = 5.57$, $P < .001$. Romanticism was the only direct predictor related to this attitude.

Participants who indicated less romanticism were more likely to believe that dating involves manipulation and shouldn't be taken seriously.

Table 5. Regressing Dating Attitudes on Individual Variables and Motives

	Dating is a Game	Men are Sex-driven	Women are Sex-objects
Predictors	β	β	β
Step 1			
Gender	-.11	.03	$-.22^{**}$
Romanticism	$-.25^{***}$	$.18^{*}$	.07
Perceived TV Realism	.14	$.17^{*}$	$.18^{*}$
Step 2			
Voyeurism	.14	$.29^{**}$	$.20^{*}$
Companionship	.15	-.17	.12
Social Learning	.15	.00	-.03
Social Interaction	-.12	.10	-.10
Passing Time	.12	-.07	-.05
Entertainment	-.06	-.12	.08
Final Model R^2	$.23^{**}$	$.26^{***}$	$.20^{a}$

Note: $N = 181$. All betas are final standardized betas on the last step of the regression.
$^{*}p < .05$. $^{**}p < .05$. $^{***}p < .001$.
$^{a}p = .06$.

Men are Sex-driven. Research question three examined the predictors associated with the attitude that men are sex-driven. The total model related to this attitude explained 26% of the variance, $F_{(9,\ 169)} = 6.43$, $P < .001$. This model included three direct predictors: perceived TV realism, romanticism, and the voyeurism motive. Participants that view reality dating shows to enjoy the sexual content, who held stronger romantic beliefs, and who believe that television programming is real were more likely to indicate they believed that men are sex-driven. Perceived realism was also an indirect predictor through the voyeurism motive. Those who believe that television is realistic were more likely to view RDTV programming to observe sexual content.

Women are Sex-objects. The last research question posed examined the predictors related to the attitude that women are sex-objects. This model was also significant, $F_{(9,\ 169)} = 4.70$, $P < .001$, and the combined effects of these variables explained 20% of the variance in participants' attitude regarding women. The direct predictors of this attitude emerged: gender, perceived realism, and the voyeurism motive. Perceived realism was also found to be an indirect predictor of this attitude through the voyeurism motive. Men who believe that television portrayals are more real and who watch RDTV programs for their sex appeal are more likely to believe that women are sexual objects whose value is based on physical appearance.

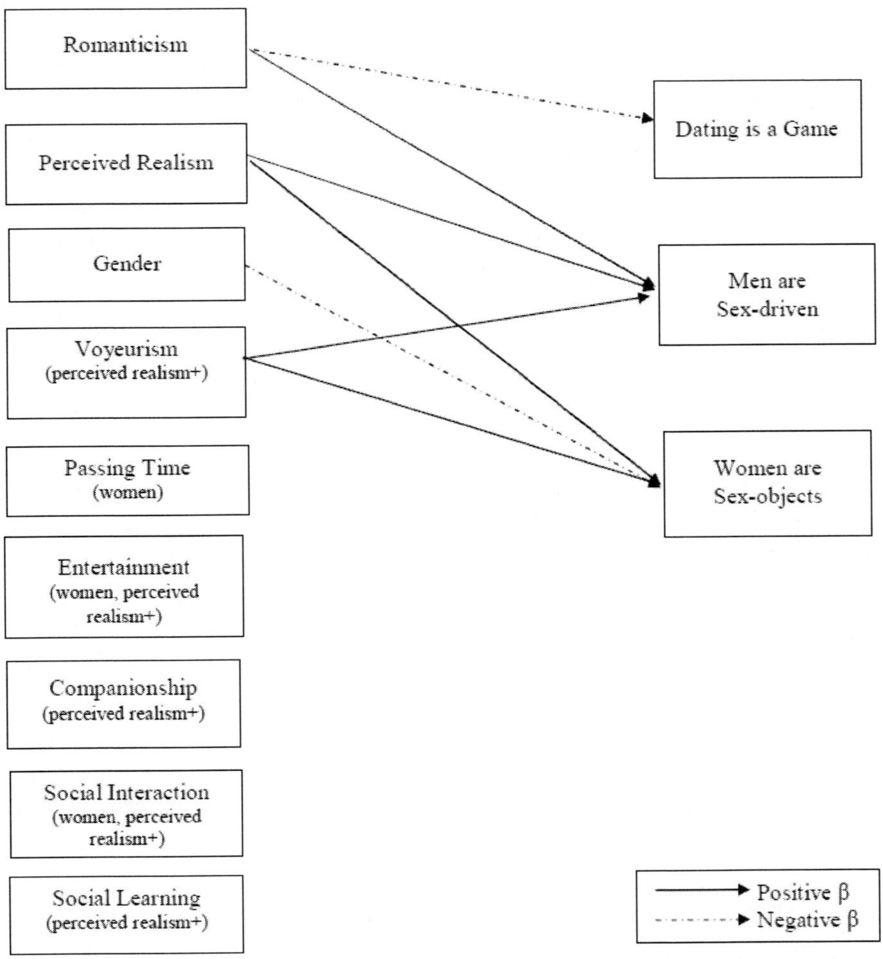

Note: Significant predictors of each RDTV motive are listed in the box in parentheses. See Table 3 for standardized beta weights related to these predictors.

Figure 2. Model of Significant Predictors of Stereotypical Dating Attitudes.

Discussion

The goal of this study was to explore the relationship between reality dating television motives and three stereotypical dating attitudes: dating is a game, men are sex-driven, and women are sex-objects by applying the uses and gratifications framework. Individual characteristics of gender, perceived realism, and romanticism were also assessed regarding their impact on motives and dating attitudes. From this data, several interesting results emerged.

First, several RDTV motives were related to holding stereotypical dating attitudes. The motives of companionship, social learning, and voyeurism were positively related to all three dating attitudes. The more you use RDTV programming to feel less lonely, to learn about others, and to view sexual content, the more likely you are to believe that dating is a game, that all men think about is sex, and that women are valued for their appearance and should expect sexual advances. Additionally, two motives were also related to the attitude that men are sex-driven, but not to the other attitudes: social interaction and entertainment. When viewers use RDTV in order to have something in common with other people and to have fun, they are more likely to believe that men are sex-driven.

According to Papacharissi (2008), different media effects will result based on whether a television viewer uses the medium for habit or to pass the time (ritualized) or more purposive means such as learning (instrumental). This study lends support for this assumption. The ritualistic motive of passing time was unrelated to any of the dating attitudes, whereas the other instrumental motives were correlated with at least one dating attitude.

Another goal of this study was to test the U&G model in relation to individual characteristics and motives as predictors of dating attitudes. Overall, the combined effects of gender, romanticism, and perceived realism with respect to RDTV motives significantly predicted each of the dating attitudes. However, specific results did emerge when exploring direct predictors.

Although romanticism did not predict any of the RDTV motives, it did emerge as a direct predictor of the attitudes dating is a game and men are sex-driven. Specifically, those who were less romantic believed that dating was a game, and those who were more romantic believed men are sex-driven. Given that romanticism encompasses the feeling that there is one true love out there for all of us and a belief in love at first sight (Sprecher & Metts, 1989), it makes sense that those who were more romantic would not believe that dating was a matter of winning or manipulation.

However, the attitude that men are sex-driven in relation to romanticism is more difficult to explain. This result may be due in part to how we view men and masculinity. Sprecher and Metts (1989) found that romanticism was positively related to the masculine gender role orientation. Television may also shape our views of masculinity.

In a study using both experimental and survey methods, Ward and Friedman (2006) found that when exposed to sexual content on television, participants were more likely to endorse sexual stereotypes and gender roles after this exposure than were participants who viewed non-sexual television clips. Survey results also showed that those who viewed more sexually-related prime-time television overall were more likely to hold these stereotypical beliefs (Ward & Friedman, 2006). It may be that these views of masculinity are embedded in our perceptions of romance, and therefore those who are more romantic may then be more drawn to reality dating programming. However, this relationship should be further explored in future research.

In addition to romanticism, perceived realism emerged as positive direct predictors of the gendered dating attitudes related to men and women. These findings have been supported by previous research regarding dating attitudes and perceive realism (Ferris et al., 2007), and indicate the importance of reality perceptions in relation to media effects. Interestingly, the motive most related to sexual content, voyeurism, was also a positive predictor of these two attitudes. It is not surprising that those who are drawn to RDTV for sexual enjoyment purposes also hold stereotypical views regarding gender and dating.

Lastly, gender was found to be a direct predictor of the belief that women are sex-objects, along with perceived realism and voyeurism. Thus, men who believe TV is more realistic and view RDTV for sexual reasons were most likely to subscribe to unhealthy attitudes regarding women. This finding is perhaps the most alarming in relation to college students populations. Given concerns over violence against women on college campuses, any negative effects related to men's perceptions of women as sex objects raises some concern.

Previous research has shown that television programming can cultivate negative attitudes toward rape and sexual assault. Kahlor and Eastin (2011) found that viewing soap operas leads to a stronger belief in rape myths, including the belief that women who dress or act in a certain way deserve to be assaulted. Future research should continue to explore how RDTV shows may cultivate other unhealthy beliefs about women and other viewing outcomes.

Study Limitations and General Conclusions

While the results do shed some light on the role of individual characteristics in relation to motives, there are some limitations embedded in the design. First, the findings are relevant only to college students. Although college students fit into the RDTV viewing demographic, future research should be conducted on non-college students to determine if any differences in motives arise. Additionally, our sample was also predominantly female. Given the potential for gender influence, future research should attempt to find a more evenly distributed gender-wise sample.

Lastly, the study was predominantly focused on heterosexual relationship dating attitudes. Although programs like the *Bachelor* and *Bachelorette* have yet to choose a homosexual primary dater, other programs like *The Millionaire Matchmaker* and *The Singles Project* have included gay and lesbian characters. Future research should further explore the dating stereotypes related to homosexual singles, as well as how motives for viewing these shows impacts those attitudes.

Overall, this study produced some interesting new findings in relation to romantic beliefs, gender, and perceived realism in relation to RDTV motives and attitudes toward dating. In particular, romanticism, perceived realism, and the motive to watch reality television for its sexual content were important predictors of stereotypical attitudes about dating. As the reality dating show genre continues to remain popular and begins to integrate more with social media, researchers should continue to explore how these shows impact viewers.

REFERENCES

The Bachelor: Season 19 Ratings (2015, March 10). *TV Series Finale.* Retrieved from http://tvseriesfinale.com/tv-show/the-bachelor-season-19-ratings-35170/.

Black, E. (2015, September 14). VH1 announces another season of everyone's favorite naked dating series! Retrieved from http://www.vh1.com/news/205761/dating-naked-gets-the-greenlight-for-season-three/.

Chapdelaine, A., Levesque, M. J., and Cuadro, R. M. (1999). Playing the dating game: Do we know whom others would like to date? *Basic and Applied Social Psychology*, 21, 139-147.

Dos Santos, K. (2015, February 23). How *The Bachelor* did the impossible: Got better than ever in season 19. Retrieved from http://www.eonline.com/news/626824/how-the-bachelor-did-the-impossible-got-better-than-ever-in-season-17.

Eggermont, S. (2006). Television viewing and adolescents' judgments of sexual request scripts: A latent growth curve analysis in early and middle adolescence. *Sex Roles*, 55, 457-468, doi:10.1007/s11199-006-9099-7.

Ferris, A. L., Smith, S. W., Greenberg, B. S., and Smith, S. L. (2007). The content of reality dating shows and viewer perceptions of dating. *Journal of Communication*, 57, 490-510. doi:10.1111/j.1460-2466.2007.00354.x.

Flomenbaum, A. (2015, March 13). 'Love,' 'hate,' 'happy,' and 'crazy': How fans reacted on Twitter to 'The Bachelor' season 19 finale. *AdWeek.* Retrieved from http://www.adweek.com/lostremote/love60.

Fogel, J., and Kovalenko, L. (2013). Reality television shows focusing on sexual relationships are associated with college students engaging in one-night stands. *Journal of Cognitive and Behavioral Psychotherapies*, 13, 321-331.

Godlewski, L. R., and Perse, E. M. (2010). Audience activity and reality television: Identification, online activity, and satisfaction. *Communication Quarterly*, 58, 148-169. doi: 10.1080/01463371003773358.

Haridakis, P. M., and Rubin, A. M. (2003). Motivation for watching television violence and viewer aggression. *Mass Communication and Society*, 6, 29-56.

Hefner, V., and Wilson, B. J. (2013). From love at first sight to soul mate: The influence of romantic ideals in popular films on young people's beliefs about relationships. *Communication Monographs*, 80, 150-175.

Hetsroni, A. (2000). Choosing a mate in television dating games: The influence of setting, culture, and gender. *Sex Roles*, 42, 83-106.

Hetsroni, A. (2012). Associations between television viewing and love styles: An interpretation using love styles. *Psychological Reports*, 110, 35-50. doi:10.2466/17.PR0.110.1.35-50.

Hetsroni, A., and Bloch, L. (1999). Choosing the right mate when everyone is watching: Cultural and sex differences in television dating games. *Communication Quarterly*, 47, 315-332.

Huff, R. M. (2006). *Reality Television.* Westport, CT: Praeger Publishers.

Kahlor, L., and Eastin, M. S. (2011). Television's role in a culture of violence toward women: A study of television viewing and the cultivation of rape myth acceptance in the United States. *Journal of Broadcasting and Electronic Media*, 55, 215-231. doi:10.1080/08838151.2011.566085.

Metcalf, M. (2015). *Top 100 Wednesday cable originals* (and network update): 9.16.2015. Retrieved from http://www.showbuzzdaily.com/articles/showbuzzdailys-top-100-wednesday-cable-originals-9-16-2015.html.

Papacharissi, Z. (2008). Uses and gratifications. In: M. Salwen, and D. Stacks (Eds.), *An integrated approach to communication theory and research* (pp. 137-152). Mahwah, N.J.: Lawrence Erlbaum Associates, Inc.

Papacharissi, Z., and Mendelson, A. L. (2007). An exploratory study of reality appeal: Uses and gratifications of reality TV shows. *Journal of Broadcasting and Electronic Media*, 51, 355-370.

Punyanunt-Carter, N. M. (2006). Love on television: Reality perception differences between men and women. *North American Journal of Psychology*, 2, 269-276.

Rivadeneyra, R., and Lebo, M. J. (2008). The association between television-viewing behaviors and adolescent dating role attitudes and behaviors. *Journal of Adolescence*, 31, 291-305. doi:10.1016/j.adolescence.2007.06.001.

Roberti, J. W. (2007). Demographic characteristics and motives of individuals viewing reality dating shows. *The Communication Review*, 10, 117-134. doi:10.1080/10714420701350403.

Rubin, A. M. (1981). An examination of television viewing motivations. *Communication Research*, 8, 141-165.

Rubin, A. M. (2009). Uses-and-gratifications perspective on media effects. In: J. Bryant and M. B. Oliver (Eds.), *Media effects: Advances in theory and research* (pp. 165-184). New York, NY: Lawrence Erlbaum.

Spangler, T. (2014, July 1). Cabler touts reality TV show as providing 'real-time' interaction between fans and cast members. *Variety*. Retrieved from http://variety.com/2014/tv/news/bravo-singles-project-will-give-fans-social-feedback-loop-into-dating-show-exclusive-1201256090/.

Sprecher, S., and Metts, S. (1989). Development of the 'Romantic beliefs scale' and examination of the effects of gender and gender-role orientation. *Journal of Social and Personal Relationships*, 6, 387-411.

Vandenbosch, L., and Eggermont, S. (2011). *Temptation Island, The Bachelor, Joe Millionaire*: A prospective cohort study on the role of romantically themed reality television in adolescent's sexual development. *Journal of Broadcasting and Electronic Media*, 55, 563-580. doi:10.1080/08838151.2011.620663.

Ward, L. M. (2002). Does television exposure affect emerging adults' attitudes and assumptions about sexual relationships? Correlational and experimental confirmation. *Journal of Youth and Adolescence*, 31, 1-15.

Ward, L. M., and Friedman, K. (2006). Using TV as a guide: Associations between television viewing and adolescents' sexual attitudes and behavior. *Journal of Research on Adolescence*, 16, 133-156.

Ward, L. M., and Rivadeneyra, R. (1999). Contributions of entertainment television to adolescent's sexual attitudes and expectations: The role of viewing amount versus viewer involvement. *Journal of Sex Research*, 36, 237-249.

Yahr, E., Moore, C., and Chow, E. (2015, May 29). How we went from 'Survivor' to more than 300 reality shows: A complete guide. *The Washington Post*. Retrieved from https://www.washingtonpost.com/graphics/entertainment/reality-tv-shows/.

Zurbriggen, E. L., and Morgan, E. M. (2006). Who wants to marry a millionaire?: Reality dating television programs, attitudes toward sex, and sexual behaviors. *Sex Roles*, 54, 1-17. doi:10.1007/s11199-005-8865-2.

In: Television and Romance
Editor: Amir Hetsroni

ISBN: 978-1-63485-077-3

Chapter 8

USING CULTIVATION THEORY TO ANALYZE COLLEGE STUDENT ATTITUDES ABOUT THE DATING PROCESS FOLLOWING EXPOSURE TO ROMANTIC FILMS

Emilio S. Hernandez and
Narissra Maria Punyanunt-Carter
Texas Tech University, Lubbock, TX, US

ABSTRACT

This cultivation study focuses on romantic cinema and questions whether individuals who are heavy viewers of romance films believe that beauty is a form of power. One-hundred and forty first-year college students were given a 32-item survey that asked if exposure to romantic films significantly affected their attitudes about dating. The results showed significant correlation between college attitudes about beauty and the frequency of viewing romantic films. The effect was even stronger among females.

The Portrayal of Romance in the Media: A Cultural Product

The mass media provide audience members with various depictions of love, while also highlighting the changing social conventions of intimate relationships (Klocker & Stanes, 2013; Von Eicken, 2014). For example, modern romantic relationships are now sometimes depicted on cinema and television as postromantic interactions with female characters becoming antagonistic and male characters acting insecure (Von Eicken, 2014). Nonetheless, the classical theme of attaining love is still culturally prevalent and central to the narrative of courtship (Von Eicken, 2014).

Arranged marriages have been a long-standing practice that has been abandoned for contemporary courtship practices (Hollander, 2004; Weisberg & Appleton, 2003). Prior to the shift towards courtship modernity, dating was a family affair, in which arrangement was often guided by family observation, but implicitly steered by economic wealth (Hetsroni, 2000). The shift towards independent mate selection was evident during the late Colonial period. With little intervention from parents, adolescents started selecting peers to date (Rothman, 1984). Boys and girls initiated romantic meetings at community events or church socials, in shops or on the street (Hetsroni, 2000; Whyte, 1992). While intimacy was not considered to be a vital trait of marital success (Murstein, 1974), it has surfaced as an important component of dating.

Love as a component of intimacy contributes to the formation of varying experiences concerning romance, which can change throughout the duration of a relationship (Ahmetoglu, Swami, & Chamorro-Premuzic, 2010). Thus, intimacy within romantic relationships is defined as "the level of commitment and positive affective, cognitive, and closeness one experiences with a partner in a reciprocal relationship" (Moss & Schwebel, 1993, p. 32). For adolescents however, ideas that help define intimacy and romance can become distorted ideals, fueled by the convenience of media content and the abundance of media messages, which can guide some young people through the process of dating, intimacy, and sex (Ward & Rivadeneyra, 1999). The contradictions about intimacy and romance adolescent boys and girls experience can be partially attributed to the persuasiveness of the mass media (Johnson & Holmes, 2009). Furthermore, this contradiction can also be attributed to the dearth of life experience adolescents have, which cannot be easily compared against media content, resulting in the belief that frequently watched programs and successful films reflect cultural norms and expectations about romance (Johnson &

Holmes, 2009). Prior research has gleamed that adolescents' exposure to sexually-oriented content heavily influences their expectations about peer sexual behavior and their ideas about permissiveness, resulting in an increase of premarital sex (Bleakley, Hennessy, & Fishbein, 2011; Bleakley, Hennessy, Fishbein, & Jordan, 2008;Varava & Quick, 2015). In a study that looked at the prevalence of adolescent sexual activity, results indicated that nearly 70% of teen boys and over half of teen girls participate in casual relationships (Manning, Giordano, & Loanmore, 2000). For some adolescent girls, sexuality is documented to occur with partners who they have dated irregularly but also with partners with whom there is little history (Elo, King, & Furstenberg, 1999; Manning, Longmore, & Giordano, 2000). Because adolescents spend more time around each other while in school, and have greater exposure to sexual relationships than previous generations had (Milbrath, Ohlson, & Eyre, 2009;Pardun, L'Engle, & Brown, 2005), adolescents now are more likely to engage in premarital sex and to believe that dating is a game. Futhermore, heavy consumption of "promiscuous media" may bring adolescents to adopt the ideals promoted by this media i.e., that women are sex objects and that men are only driven by sex (Kaloff, 1999; Ward, 2002; Zurbriggen & Morgan, 2006). This process is the basic principle of cultivation theory: the more you view recurring media portrayals, the more likely you are to internalize them (Gerbner, Gross, Morgan, & Signorielli, 1994).

Purpose of the Study

Cultivation theory has endured some criticism for failing to acknowledge genre-specific contexts outside of television (Arendt, 2010; Good, 2008). According to cultivation theory, stories that disseminate recurrent or similar messages can alter a viewer's perception about social reality. This is why the *narrative* is considered to be a significant genre-specific context where stories are placed into categories that convey a common theme. The intensity of the narrative experience and the lack of adequate counterarguments strengthens the effectiveness of the narrative by helping to reshape viewer attitudes and beliefs (Bilandzic & Buzzell, 2008). This process is known as narrative transportation (Green & Brock, 2000). Narrative transportation is concerned with the interaction audience members have with the narrative as it unfolds on the screen (Shrum, Lee, Brroughs, & Rindfleisch, 2011). Romantic media is an ideal exemplar of effective narrative in an effort to maintain a heightened state of emotional arousal viewers may begin making trait judgements about characters and start cultivating attitudes (perhaps based in part on the extent to which characters share the same values as the viewers) (Shrum et al., 2011). Because these second order judgments occur spontaneously and during the viewing process, they can elicit a greater effect of cultivation among heavy media consumers (Shrum & Bischak, 2001; Shrum et al., 2011). Thus, by considering the romantic movie genre as the medium of choice for this study, we hope to learn about the role of beauty in the formation of an intimate relationships because beauty is highlighted in romantic narratives. We view this belief as a combination of *first order cultivation* (reality-based estimates that align with media depictions) and *second order cultivation* (beliefs and values which are guided by media messages).

Cultivation Theory Summary

George Gerbner, Larry Gross, Michael Morgan, and Nancy Signorielli have developed Cultivation Theory, which is centered on the belief that the consistency and ubiquity of media portrayals, in particular television, have made it the primary source of information about the social world (Gerbner, Gross, Morgan, & Signorielli, 1994). In its simplest explanation, this theory asserts that the direct experience of reality will be replaced with the indirect experience of watching television, and as a result, influence the formation of social beliefs for individuals (Gerbner et al., 1980). This stems from Gerbner's position that the industrial revolution of technology creates symbolic environments that reflect the institutions – the function and structure of the decision-making pressures, organizational forms, and power relations that disseminate particular messages (Gerbner, 1998; 1999). Gerbner further posits that the mass production of messages and the rapid distribution thereof likewise contribute to the collective consciousness (shared meaning) that is formed and imposed on social relationships (Gerbner, 1998). Shared meaning refers to a unanimously-held conception individuals have about reality that is based on the amount of television a person views (Gerbner, 1998). It is these frequent messages that serve as the infrastructure for symbolic environments that aligns with cultivation theory that serves as the expressive voice for culture and society (Gerbner, 1998; 1999). However, the degree to which these symbolic environments effects viewers is predicated on the mainstreaming and resonance effects. Mainstreaming is moderated by direct life experience, which is the homogenization of viewer perceptions, and usually is characteristic of low viewers of media (Gerbner, Gross, Morgan &

Signorielli, 1980). Unlike mainstreaming, however, resonance predicts an interaction will occur when life experience and the experience portrayed through media appear congruent for the viewer (Gerbner et al. 1994). Thus, it is the sharing of the institutional and social dynamics that construct symbolic environments through message dissemination, which is the outcome of the cultivation process (Gerbner, 1998; 1999).

Previous research indicates the television is a socially transformative device because of its ability to guide cultural and social perceptions, and unify groups of people through message dissemination (Gerbner, 1998). This view aligns with Gerbner's position that suggests heavy television viewing results in inaccurate depictions of reality (initially crime and law enforcement), which creates an exaggerated perception about danger, mistrust, and victimization (Signorielli & Gerbner, 1995). Furthermore, this highlights the limitless capability of media, which grabs the attention of nations, and mobilizes the behavior of large groups of individuals – proposing the television is the "modern propaganda machine" that creates a cohesive environment of shared meaning among audiences (Gerbner, 1998). If television is viewed for long periods of time, the symbolic environment that is created by media will likely cultivate a perception that suggests the world is how media portrays it as (Mutz & Nir, 2010). For high viewers of television, this is problematic because recurrent messages that occur frequently repeat a patterned message that increases the likelihood that these viewers will have a strong effect of cultivation (Gerbner, 1998).

LITERATURE REVIEW

Viewer Emotions

Romantic content can be dramatic, humoristic or tragic. Therefore, consumers of romantic narratives can experience a wide range of emotions (Harris, Hoekstra, Scott, Sanborn, Dodds & Brandenburg, 2004). Examining the types of emotional outcomes that viewers experience, found viewers' sentiment after watching movies with a tragic end helps to reinforce the moral value of empathy among women (Oliver, 1993). Furthermore, tragic entertainment elicits pleasurable emotions such as compassion and tenderness (Oliver, 1993). One example of tragic content that may elicit romantic behavior is horror films (Harris, et al., 2004). Research shows that watching films of this genre may induce both males and females to sense interdependence towards their partners positive affect for both males and females is derived from the sensation of interdependence both partners exhibit towards one another (Harris et al., 2004). As a rule of thumb, gratification is cited as the primary reason why people choose to watch movies (Bartsch, 2012). The kind of gratification from media consumption is closely related to the types of emotion a person experiences while (or right after) viewing (Bartsch, 2012). The emotion of love could be one of the reasons why individuals choose to view romance films (Grodal, 2009; Schulze, 1986).

Love is perceived to be an invincible force that triumphs over gender and social status (Abbott & Jermyn, 2009). In cinematic terms, romantic love is considered to be a "utopian force that brings together (usually a heterosexual couple) who are destined to be together" (Abbott & Jermyn, 2009, p. 137). Since the implicit aim of the romantic genre is to guide viewers on relationships, it is conceivable to suppose that the film industry creates a "magical perception" of love that is transformative and extends into reality (Abbott & Jermyn, 2009). By projecting the feeling of love onto the silver-screen, the film industry capitalizes on this emotion by suggesting this sensation is a powerful feeling a person will experience (Grodal, 2009). The film industry accomplishes this by featuring narratives that for example, promote a kiss is able to remedy conflicts (Grodal, 2004; 2009); Urang, 2006). Love is presented as something absolute because the negotiation thereof is a powerful force that can either cause unity or tragedy – highlighting the permanence and strength of love (Grodal, 2009). Furthermore, the cinematic genre of romance is appealing because romantic content is familiar and contributes to the feeling of nostalgia that allows the past to be remembered (Hoffmann, 2004; Person, 1988). Individuals who choose this genre most often do so because this context is not contingent on novelty, but relies heavily on a near perfect prediction of a romantic outcome that satisfies a need for the viewer (Hoffman, 2004; Simon et al., 1992). In addition, romantic motion pictures are appealing because they depict a romantic journey that viewers can identify with, thus merging identity & myth together so that a film can be perceived as real and symbolic (Abbott & Jermyn, 2009). This allows the social and psychological conditions of a successful romance to be experienced without the burden of moral trappings (Ellis, 1990; Light, 1984; Simon et al., 1992). For instance, Indian romantic films portray the tenet of romance as a sanitary unproblematic environment where love and romance are a pure and perfect journey of internalization a person endures to obtain love (Abbott & Jermyn, 2009). Within the United States, a paradox exists for the adult viewer who remembers high school as possibly depressing and dreadful, but the romantic teen is able to transform his or her negative experience into a fantasy that is metaphorically situated on the prom as the fulcrum for romance (Radner, 2011). This allows the adult viewer to relive their high school experience through a frame of wish fulfillment, while understanding that pain and torment are necessary to obtain love (Radner, 2011). The comfortable suspension between realism and romance, coupled with the entertainment factor, conjures the feeling of love, which is one reason why romantic films are appealing (Grodal, 2004; Urang, 2006).

For some, intimate relationships are hinged on the idealistic expectations individuals believe are vital components for romance (Hefner, 2008). People that believe this romantic paradigm generally have unprecedented access to messages that convey particular social standpoints, suggesting this form of media exposure contributes to the second order effect of cultivation (Bachmen & Illouz, 1996). Since messages serve as the symbolic infrastructure for culture (Shrum & Bischak, 2001), it is possible to conclude that certain standpoints, are cultural messages that have the potential to shape individual perceptions (Bachmen & Illouz, 1996). Since perceptual change is socially contingent on viewers encoding & decoding information gathered from media (Bilandzic & Rössler, 2004), cultivation theory posits individuals begin to acquire knowledge, and start to form perceptions from stored information (Shrum & Bischak, 2001).

Romantic content focuses on the unsatisfied idealistic elements of relationships people are in search of by creating a fictional image of a romantic couple (Harrington & Bielby, 1991). With the integration of the traditional components of commitment, intimacy, and trust, media are able to focus on the features of romance that are appealing to individuals (Harrington & Bielby, 1991). In this case, the media are able to focus on the myth of love (Harrington & Bielby, 1991). This idealistic conception of popular romance reinforces patriarchal values by promoting heterosexuality, marriage, and monogamy are components indicative of a successful relationship (Jarvis, 1999).

Hypotheses

Research indicates that media portrayals of romance that mirror actual relationships may cultivate a belief among viewers that suggest the components of romance are possible to obtain (Zhou & Zhu, 2004). Since media consumption is a central activity for many young people, they may use the media as a source of exemplars for retrieval concerning matters of romance

(Eggermont, 2004), as cultivation theory would predict (Gerbner et al., 1980; Morgan & Signorielli, 1990), we present the following hypothesis:

H1: The frequency of watching romance films will alter college students dating beliefs.

Since women are known as more prone to romantic TV shows (Segrain & Nabi, 2002), we also hypothesize that:

H1: Female students will be more highly influenced than male students.

METHOD

Participants

Participants for this research study included N = 248 undergraduate students enrolled at a large southwestern university. Surveys were administered to students enrolled in an interpersonal communication course, and in a communication, civility, and ethics course. Out of the original participants, those who selected other class designations other than first-year status were removed from the sample population. Limiting this survey to first-year college students is based on college transition research that indicates many high school students generally lack cultural, peer, and networking support, as well as being deficient in content knowledge, skills, or habit of mind needed to perform college work successfully (Venezia & Jaeger, 2013). Coupled with this is the transition from high school into college, which for some can be a difficult event because of inadequate time to have developed sufficient emotional and life experience (Gall, Evans, & BeUerose, 2003; Park, Edmondson, & Lee, 2012). Because this transition involves many life-changing events, some first-year students may find it difficult to establish new friendships and successfully cope with coursework expectations (Park et al., 2012). For those individuals who cope poorly with the transition into college, this can result in students becoming isolated and avoid using college services meant to help students become successful (Lopez & Gormey, 2002). This has reduced the number of respondents to 86 (61%) females and 54 (39%) males n = 140. The vast majority of participants 137 (97.86%) reported to be 18-24 years of age and 3 (2.14%) reported to be 17 years of age. The ethnicity breakdown is as follows: 5 3.5%; African-American, 7 5%; Asian, 5 3.5%;

Caucasian, 97 69%; Hispanic, and 26 19%; Native American. Of the 140, 63 were high viewers of romantic films, 47 moderate viewers, and 30 were low viewers of romantic films.

Instrument Description

A modification of Cobb, Larson, and Watson's (2003) Attitudes about Romance and Mate Selection Scale was used for this study. Originally, the survey was developed to measure constraining beliefs about mate selection, but for this research, the survey has been modified to measure if exposure to romantic films cultivates an attitude of romantic idealism among college students. The self-report format has been deemed appropriate for this research study because (Free, Boltan, & Kreps, 2000), the self-report method allows respondents to rate their attitudes, beliefs, and values, which is a reliable way to learn how people feel about a particular topic or phenomenon being measured. The survey includes five attitude beliefs college students may potentially have when viewing romantic films, which are collected from the literature and modified from the original research. The five attitudes are: cohabitation, beauty is power, dyadic power negotiation, idealization, and love. A seven point scale was used to code the answers ranging from 1 (very strongly disagree) to 7 (very strongly agree). Cronbach alpha of the scale was .91. Cronbach –alpha coefficients of the subscales were: cohabitation (α = .88), beauty is power (α = .73), dyadic power negotiation (α = .92), idealization (α = .82), and love (α = .85). Hence, the scales were reliable. The survey featured five demographic questions: age, class designation, income, sex, and race. The survey consists of 32 questions that deal with the five attitudes students have by devoting five questions to each belief, as well as focusing on the hypothesis proposed. Appendix A features the survey questionnaire.

RESULTS

A preliminary factor analysis was conducted to verify the relevant factor scales. A principal components analysis with varimax rotation was conducted. Results yielded five factors that accounted for 3.11% of the total variance. Eigenvalues of the factors in the 32-item scale were between .28 and. 93. The factor cohabitation accounted for 4.42% of the variance and had values from .88 to .94. The factor for beauty is power had values between .71 and .82 and accounted for 2.22% of the variance. The factor for dyadic power negotiation has values between .55 and .69 and accounted for 1.75% of the variance. The factor for idealization accounted for 1.47% of the variance and had values that ranged from. 55 to .73. The factor for love had valued from .43 to .77 and accounted for 1.14%. The Skree Plot (see Appendix B) indicates that four eigenvalues are significant: cohabitation (factor 1), beauty (factor 2), power (factor 3), love (factor 4).

In order to test H1, we conducted an ANOVA where the independent variable is the amount of time devoted to watching romantic films and the dependent variables were the five subscales. Gender was used as a control variable. There was a difference between those participants (M = 3.40, SD = .82) who watched significantly more romantic films (n = 63, t = 5.41, p < .001) than those who reported that they were low/non-viewers of romantic films (n = 30, M = 3.97, SD = .62). Significant differences among the two groups (high viewers and low viewers) in terms of the subscales revealed that love (4, 139) = 6.12, p < .01 and beauty is power (4, 134) = 76.22, p < .01, and idealization (4, 138 = 43.21, p < .01 were significant. A Tukey post hoc tests revealed that high viewers (M = 4.88, SD = .69) of romantic films scored higher (p <. 01) on average than low viewers of romantic films (M = 3.70, SD = .78) on the three significant items. H1 is supported.

To see if there was a difference between males and females on each of the subscales and test H2, we ran an independent samples t test. There were three significant gender differences: love (n = 139, t = -2.22, p < .01); idealization (n = 138, t = -3.14, p <. 01), and beauty is power (n = 134, t = -3.38, p < .01) with women rating those subscales higher than men. This confirms H2.

DISCUSSION

Romantic films have garnered appeal from viewers because research indicates people want to observe the success of romantic relationships despite challenges (Hefner & Wilson, 2013). Romantic content has become a mainstay in popular culture because adolescents in particular, are in search of information that teaches dating behavior (Hefner & Wilson, 2013; Zurbriggen

& Morgan, 2006). Thus, concerning the present study, establishing a correlation between beauty and power as a component of dating and watching romantic films did yield confirming values. The data confirm not only the effect of film viewing but also gender differences with women exhibiting stronger belief in traditional relationship ideals.

Why does this happen? Frequent movie goers may compare the fictional message against their personal life, and make sound judgments that align with the film description (Milkie, 1999). Even if these viewers realize that that these descriptions are fictionalized and made merely for entertainment purposes, they may still find them appealing (Ameri, 2011). As for the gender differences - female college students may be more susceptible than males to endrose traditional relationship ideals either because they watch more romantic films or because of gender differences in cultural upbringing (Haferkamp, 1999).

RESEARCH LIMITATIONS AND CONCLUSION

A clear limitation of this research study is the single class of students surveyed. Original conceptions have kept the survey limited to first-year students, so generalization options are limited. Additionally, narrowing the scope of the survey to focus on films of heavy romance leave the topic somewhat open to confusion among some participants because it was not clear if a particular film was being evaluated. While this strategy is typical of cultivation research, the problem is not immediately dismissed (Shrum et al., 2011). Moreover, research indicates that in a survey such as ours participants may have low incentive to answer the questions (Shrum, 2007).

In sum, the aim of this study was to provide an additional context to cultivation theory and determine whether exposure to romance films influences college student attitudes about dating. The results indicate that students are influenced in general and that females are particularly impacted by exposure to romantic film content. Media literacy programs regarding romantic media may be beneficial and necessary for college students in general and for female students in particular.

Appendix A

Survey instrument

Directions:

The purpose of this survey is to measure if exposure to romantic films alters college student attitudes about dating. Please note that some questions may appear repetitive, but this is necessary for the research. Know (All participants and surveys). This survey consists of 32 questions and should take no more then ten minutes to complete. If at any time you feel that your participation will cause you harm, or if you simply wish to no longer participate in this research, just return your survey to the researcher upon collection. There is no wrong answer, so please do your best to answer all the questions provided. All the answers will remain anonymous. Finally, your participation is voluntary and appreciated very much.

Please circle your response to the following 5 demographic questions.

1 How old are you?

17 18-24 25-30 31-35 36-40 41-45 46 and older

2 What is your current class designation?

First-year (freshman) Sophomore Junior Senior Graduate/ professional Non-traditional Other Please explain__________

3 What is your annual income?

$0-5000 $6000-$10,000 $11,000-20,000 21,000-30,000 31,000-40,000 41,000-50,000 51,000 and above

4 What is your sex?

Female Male

5 What is your racial background?

African-American Asian Black Caucasian/white Hispanic/Latino Native American Other Please explain__________

Please circle your response to the following two questions that ask how often you view romantic films, and then average your scores for both questions. This is important for later data analysis.

6 How many hours a day do you view movies that focus on romance?

0-1 hours 1-2 hours 2-3 hours 3-4 hours 4-5 hours 5-6 hours 6 or more hours

7 How many times a week do you view movies that focus on romance?

0-1 times 1-2 times 2-3 times 3-4 times 4-5 times 5-6 times 6 or more times

Please write average score here___

Please write your response in the space provided that best reflects your attitude based on your exposure to romantic films. Where 1= very strongly disagree, 2 = strongly disagree, 3 = disagree, 4 = undecided, 5 = agree, 6 = strongly agree, and 7 = very strongly agree with each statement.

8. ___ After viewing romantic films, I believe it is a good idea to live with a dating partner.

9. ___ After viewing romantic films, I believe living with a dating partner is a good way to test if the relationship will last.

10. ___ After viewing romantic films, I believe living with a dating partner will minimize the chance for divorce.

11. ___ After viewing romantic films, I believe living with a dating partner will contribute to a happy dating relationship.

12. ___ After viewing romantic films, I believe sex is an acceptable component of the dating process.

13. ___ After viewing romantic films, I believe that a dating partner should be very physically attractive without exception.

14. ___ After viewing romantic films, I believe physical attractiveness can be used for personal gain.

15. ___ After viewing romantic films, my attitude about my physical appearance is more critical.

16. ___ After viewing romantic films, physical attractiveness is something I evaluate a potential dating partner by.

17. ___ After viewing romantic films, I believe physical attractiveness is not important when participating in the dating process.

18. ___ My attitude about dating becomes more negative after viewing romantic films.

19. ___ After viewing romantic films, I develop an attitude of acceptance towards having sex when participating in the dating process.

20. ___ After viewing romantic films, I am more self-confident about my physical attractiveness and use this to my advantage in the dating process.

21. ___ After viewing romantic films, I develop an attitude of conquer and conquest concerning the dating process.

22. ___ After viewing romantic films, I develop an attitude of submission and subservience concerning the dating process.

23. ___ I consider myself to be a very romantic person.

24. ___ I believe that some romantic comedies offer a realistic image of dating relationships.

25. ___ After viewing romantic films, I believe that romantic happy endings can occur in real life.

26. ___ After viewing romantic films, my outlook on the dating process is poor because there is no ideal partner for me.

27. ___ After viewing romantic films, I have a tendency to believe the idealistic messages shared about the dating process.

28. ___ After viewing romantic films, I believe the feeling of love should be enough to sustain a healthy dating relationship.

29. ___ After viewing romantic films, I believe there is only one person I should be with.

30. ___ After viewing romantic films, I believe love is not essential for casual dating.

31. ___ I believe after viewing romantic films love is overrated and makes dating difficult.

32. ___ I believe that actual love and romance should mirror romantic films.

APPENDIX B

Catell's Test (Scree Plot)

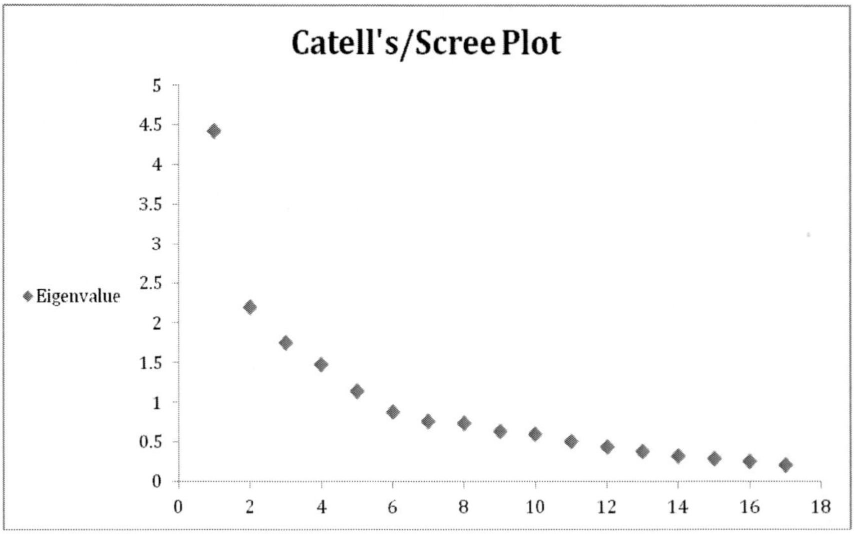

REFERENCES

Abbotte, S., and Jermyn, D. (2009). *Falling in love again: Romantic comedy in contemporary cinema.* New, York, NY: I.B. Tauris and Company Ltd.

Ahmetoglu, G., Swami, V., and Chamorro-Premuzic, T. (2010). The relationship between dimensions of love, personality, and relationship length. *Archives of Sexual Behavior*, 39(5), 1181-1190.

Alexander, A. (1985). Adolescents' soap opera viewing and relational perceptions. *Journal of Broadcasting and Electronic Media*, 29(3), 295-308.f.

Ameri, A. (2011). Imaginary placements: The other space of cinema. *Journal Of Aesthetics and Art Criticism*, 69(1), 81-91.

Appel, M. (2008). Fictional narratives cultivate just-world beliefs. *Journal of Communication*, 58(1), 62-83.

Arendt, F. (2010. Cultivation effects of a newspaper on reality estimates and explicit and implicit attitudes. *Journal of Media Psychology: Theories, Methods, and Applications*, 22(4), 147-159. Bachen, C. M., and Illouz, E.

(1996). Imagining romance: Young people's cultural models of romance and love. *Critical Studies in Mass Communication*, 13(4).

Bartsch, A. (2012). Emotional gratification in entertainment experience. Why viewers of movies and television Series find it rewarding to experience emotions. *Media Psychology*, 15, 267-302.

Becker, C. (2009). Defamiliarizing and refamiliarizing film and television texts. *Cinema Journal*, 48(3), 90-94.

Bilandzic, H., and Busselle, R. W. (2008). Transportation and transportability in the cultivation of genre-consistent attitudes and estimates. *Journal of Communication*, 58, 508-529.

Bilandzic, H., and Rössler, P. (2004). Implications of genre-specific cultivation effects: The gratification/cultivation model. *European Journal of Communication Research*, 29, 295-326.

Bleakley, A., Hennessy, M., and Fishbein, M. (2011). A model of adolescents' seeking of sexual content in their media choices. *Journal of Sex Research*, 48(4), 309-315.

Bleakley, A., Hennessy, M., Fishbein, M., and Jordan, A. (2008). It works both ways: The relationship between exposure to sexual content in the media and adolescent sexual behavior. *Media Psychology*, 11, 443-461.

Bumsub, J., and Seongjung, J. (2010). The impact of Korean television drama viewership on the social perceptions of single life and having fewer children in married life. *Asian Journal of Communication*, 20(1), 17-32.

Chung, J., and Bissell, K. (2010). Communication theory and mental image. *Paper presented at The International Communication Association Conference*; annual meeting.

Cobb, N. P., Larson, T. H., and Watson, W. T. (2003). Development of the attitudes about romance and mate selection scale. *Family Relations*, 52(3), 222-231.

Cohen, J., and Weimann, G. (2000). Cultivation revisited: Some genres have some effects on some viewers. *Communication Reports*, 13(2), 99-114.

Eggermont, S. (2004). Television viewing, perceived similarity, and adolescents' expectations of a romantic partner. *Journal of Broadcasting and Electronic Media*, 48(2), 244-265.

Ellis, K. (1990). Fatal attraction or the post-modern Prometheus. *Journal of Sex Research*, 27(1), 111-121.

Elo, I. T., King, R. B., and Ferstenberg, F. F. (1999). Adolescent females: Their sexual partners and the fathers of their children. *Journal of Marriage and the Family*, 61, 74-84.

Escobar-Chavez, S. L., and Anderson, G. A. (2008). Media and risky behaviors. *Children and Electronic Media*, 18(1), 147-180.

Gall, T. L., Evans, D. R., and BeUerose, S. (2000). Transition to first-year university: Patterns of change in adjustment across life domains and time. *Journal of Social and Clinical Psychology*, 19(4), 544-567.

Gerbner, G. (1998). Cultivation analysis: An overview. *Mass Communication and Society*, 7(3/4), 175-194.

Gerbner, G. (1999). The stories we tell. *Peace Review*, 11(1), 9.

Gerbner, G., Gross, L, Morgan, M., and Signorielli, N. (1980). The mainstreaming of America: Violence profile No. 11. *Journal of Communication*, 30(3), 10-29.

Gerbner, G., Gross, L, Morgan, M., and Signorielli, N. (1984). Political correlates of television viewing. *Public Opinion Quarterly*, 48(1B), 283-300.

Gerbner, G., Gross, L, Morgan, M., and Signorielli, N. (1994). Growing-up with television: The cultivation prospective. In: J. Bryant, and D. Zillman (Eds.), *Media Effects: Advances in Theory and Research* (pp. 17-41).

Green, M. C., and Brock, T. C. (2000). The role of transportation in the persuasiveness of public narratives. *Journal Of Personality and Social Psychology*, 79(5), 701-721.

Good, J. E. (2009). The cultivation, mainstreaming, and cognitive processing of environmentalists watching television. *Environmental Communication,* 3(3), 279-297.

Grodal, T. (2004). Love and desire in the cinema. *Cinema Journal*, 43(2), 26-46.

Grodal, T. (2009). *Embodied visions.* New York, NY: Oxford University Press Inc.

Haferkamp, C. J. (1999). Beliefs about relationships in relation to television viewing, soap opera viewing, and self-monitoring. *Current Psychology*, 18 (2), 193-204.

Harrington, C., and Bilbey, D. D. (1991). The mythology of modern love: Representations of love in the 1980s. *Journal of Popular Culture*, 24(4), 129-144.

Harris, R., Hoekstra, S. J., Scott, C. L., Sanborn, F. W., and Dodds, L. A. (2004). Autobiographical memories for seeing romantic movies on a date: Romance is not just for women. *Media Psychology*, 6(3), 257-284.

Hawkins, R. P., and Pingree, S. (1980). Some processes in the communication effect. *Communication Research*, 7, 193-226.

Hefner, V. (2008). Romantic relationship ideal in mass media: An explication of a construct. Confrence papers; National Communication Association.

Hefner, V., and Wilson, B. J. (2013). From love at first sight to soul mate: The influence of romantic ideals in popular films on young people's beliefs about relationships. *Communication Monographs*, 80(2), 150-175.

Hendricks, A. (2002). Examining the effects of hegemonic depictions of female bodies on television: A call for theory and programmatic research. *Critical Studies in Media Communication*, 19(1), 106.

Hetsroni, A. (2000). Choosing a mate in television dating games: The influence of setting, culture, and gender. *Sex Roles*, 42(1-2).

Hetsroni, A. (2008). Overrepresented topics, underrepresented topics, and the communication effect. *Communication Research Reports*, 25(3), 200-210.

Hetsroni, A. (2010). When the wind changes direction: The impact of content shift on the cultivation effect. *European Journal of Communication Research*, 35(4), 439-460.

Hetsroni, A., and Tukachinsky, R. H. (2006). Television-world estimates, real-world estimates, and television viewing: A new scheme for cultivation. *Journal of Communication*, 56(1), 133-156.

Hoffman, D., Novak, T. P., and Venkatesh, A. L. (2004). Has the internet become indespensible? Empirical findings and model development. Communications of the ACM.

Holland, M. N. (2008). Spider-man? Sure! The neuroscience of suspending disbelief. *Interdisciplinary science reviews*, 33(4), 312-320.

Hollander, P. (2004). The counterculture of the heart. *Society*, 41(2), 69-77.

Holmes, B. (2007). In search of my one-and-only: Romance-related media and beliefs in romantic relationship destiny. *Electronic journal of Communication*, 17(3-4).

Jackson, C., Brown, J. D., and Pardun, C. J. (2008). A tv in the bedroom: Implications for viewing habits and risk behaviors during early adolescence. *Journal of Broadcasting and Electronic Media*, 52(3), 349-367.

Jarvis, C. (1999). Love changes everything: The transformative potential of popular romantic fiction. *Studies in the Education of Adults*, 31(2), 109-122.

Johnson, K. R., and Holmes, B. M. (2009). Contradictory messages: A content analysis of Hollywood-produced romantic comedy feature films. *Communication Quarterly*, 57(3), 352-373.

Kaloff, L. (1999). The effects of gender and music video imagery on sexual attitudes. *Journal of Social Psychology*. 139, 378-385.

Klocker, N., and Stanes, E. (2013). Reel love across ethnic boundaries? the extent and significance of inter-ethnic intimacy in australian cinema. *Ethnic and Racial Studies*, 36(12), 2035-2054.

Kwan, S., and Trautner, M. (2009). Beauty work: Individual and institutional rewards, the reproduction of gender, and questions of agency. *Sociology Compass*, 3(1), 49-71.

Light, A. (1984). Returning to manderley: Romance female fiction and class. *Feminist Review*, 16, 7-25.

Lopez, F. G., and Gormley, B. (2002). Stability and change in adult attachment style over the first-year college transition: Relations to self-confidence, coping, and distress patterns. *Journal Of Counseling Psychology*, 49(3), 355.

Manning, W. D., Giordano, P. C., and Longmore, M. A. (2006). Hooking up: The relationship contexts of nonrelationship sex. *Journal of Adolescent Research.*

Manning, W. D., Longmore, M. A., and Giordano, P. C. (2000). The relationship context of contraceptive use at first intercourse. *Family Planning Prospectives*, 32, 104-110.

Milbrath, C., Ohlson, B., and Eyre, S. (2009). Analyzing cultural models in adolescent accounts of romantic relationships. *Journal of Research on Adolescents*, 19(2), 313-352.

Milkie, M. A. (1999). Social comparisons, reflected appraisals, and mass media: The impact of pervasive beauty images on black and white girls' self-concepts. *Social Psychology Quarterly*, 62(2), 190-210.

Morgan, M., and Shanahan, J. (1997). Two decades of cultivation research: An appraisal and meta-analysis. *Communication Yearbook*, 20, 1-45.

Morgan, M., and Shanahan, J. (2010). A state of cultivation. *Journal of Broadcasting and Electronic Media*, 54(2), 337-355.

Morgan, M., and Signorielli, N. (1990). Cultivation analysis: Conceptulization and methodology. In: Signorielli, N., and Morgan, M (Eds.), *Cultivation Analysis: New Directions in Media Effects Research* (pp. 13-34).

Moss, B. F., and Schwebel, A. I. (1993). Defining intimacy in romantic relationships. *Family Relations*, 42(1), 31-37.

Murstein, B. I. (1974). *Love, sex, and marriage through the ages*. New York: Springer.

Mutz, D. C., and Nir, L. (2010). Not necessarily the news: Does fictional television influence real-world policy preferences? *Mass Communication and Society*, 13(2), 196-217.

Newcomb, H., and Hersch, P. M. (1983). Television as a cultural forum. *Quarterly Review of Film Studies*, 8(3), 561-573.

Oliver, M. B. (1993). Exploring the paradox of the enjoyment of sad films. *Hu- man Communication Research*, 19, 315-342.

Oliver, M. B., and Bartsch, A. (2010). Appreciation as audience response: Exploring entertainment gratifications beyond hedonism. *Human Communication Research*, 36, 53-81.

Park, C., Edmondson, D., and Lee, J. (2012). Development of self-regulation abilities as predictors of psychological adjustment across the first year of college. *Journal Of Adult Development*, 19(1), 40-49.

Pardun, C. J. (2002). Romancing the script: identifying the romantic agenda in top-grossing movies. In: J.D. Brown, J.R. Steele, and K. Walsh-Childers (Eds.), *Sexual teens, sexual media: Investigating media's influence on adolescent sexuality* (pp. 211-225).

Pardun, C. J., L'Engle, K. L. and Brown, J. D. (2005). Linking exposure to outcomes: Early adolescents' consumption of sexual content in six media. *Mass Communication in Society*, 8(2), 75-91.

Person, E. S. (1988). *Dreams of love and fateful encounters*. Markahm, Onterio: Pengluin Books Canada Ltd. 639-661.

Radner, H. (2011). *Neofeminists Cinema: Girly films, chick flicks, and consumer culture.* New York, NY: Rutledge.

Rothman, E. K. (1984). *Hands and hearts: A history of courtship in America.* New York: Basic Books.

Scanlon, J. (2007). "If my husband calls I'm not here": The beauty parlor as real and representational female space. *Feminist Studies*, 33(2), 308-334.

Schulz, L. J. (1986). Getting physical: Text/context/reading and the made-for-television-movie. *Cinema Journal*, 25(2), 35-50.

Segrin, C., and Nabi, R. L. (2002). Does television viewing cultivate unrealistic expectations about marriage? *Journal of Communication*, 52 (2), 247-263.

Signorielli, N., and Gerbner, G. (1995). Violence on television: The cultural indicators project. *Journal of Broadcasting and Electronic Media*, 39(2), 278.

Signorielli, N. (1997). A content analysis: A reflection of girls in the media. *Children now and the Kaiser Family Foundation.*

Simon, R. W., Eder, D., and Evans, C. (1992). The development of feeling norms underlying romantic love among adolescent females. *Social Psychology Quarterly*, 55(1), 29-46.

Shapiro, J., and Rucker, L. (2004). The Don Quixote effect: Why going to the movies can help develop empathy and altruism in medical students and residents. *Families, Systems and Health: The Journal Of Collaborative Family Healthcare*, 22(4), 445-452.

Shrum, L. J., and Bischak, V. (2001). Testing moderators of the cultivation effect. *Human Communication Research*, 27(2).

Shrum, L. J. (1995). Assessing the social influence of television: A social cognition perspective. *Communication Research*, 22(4), 402-429.

Shrum, L. J. (2007). The implications of survey method for measuring cultivation effects. *Human Communication Research*, 33(1), 64-80.

Shrum, L. J., Burroughs, J. E., and Rinfleisch, A. (2005). Television's cultivation of material values. *Journal of Consumer Research*, 32(3), 473-479.

Shrum, L. J., Lee, J., Burroughs, J. E., and Rinfleisch, A. (2011). An online process of second-order cultivation effects: How television cultivates materialism and its consequences for life satisfaction. *Human Communication Research*, 37(1), 34-57.

Urang, J. G. (2006). Realism and romance in the eastern german cinema. *Film History*, 18(1), 88-103.

Ward, L. M. (1995). Talking about sex: Common themes about sexuality in the prime-time television programs children and adolescents view most. *Journal of Youth and Adolescents*, 24, 595-615.

Ward, L., and Friedman, K. (2006). Using tv as a guide: Associations between television viewing and adolescents' sexual attitudes and behavior. *Research on Adolescents*, 16(1), 133-156.

Ward, L. M., and Rivadeneyra, R. (1999). Contributions of entertainment television to adolescents' sexual attitudes and expectations: The role of viewing amount versus viewer involvement. *Journal of Sex Research*, 36 (3), 237-249.

Whyte, M. K. (1992). Choosing mates: The amaerican way. *Society*, 29(3), 71-77.

Venezia, A., and Jaeger, L. (2013). Transitions from high school to college. *Future Of Children*, 117-136.

Von Eicken, V. (2014). Masculinity in Crisis? - What a Man (2011) and the German 'relationship comedy.' *Journal Of Contemporary Film*, 12(1/2), 79-96.

Zhou, S., and Zhu, H. (2004). Perception of romanticism and the ideal spouse among Chinese youth. *Asian Journal of Communication*, 14(1), 22-37.

Zimmer-Gembeck, M. J., Siebenbruner, J., and Collins, W. A. (2004). A prospective study of intra-individual and peer influences on adolescents' heterosexual romantic and sexual behavior. *Archives of Sexual Behavior*, 33(4), 381-394.

Zubriggen, E. L. and Morgan, E. M. (2006). Who wants to marry a millionaire? Reality dating television programs, attitudes towards sex, and sexual behaviors. *Sex Roles*. 54(1-2).

In: Television and Romance
Editor: Amir Hetsroni

ISBN: 978-1-63485-077-3

Chapter 9

TELEVISION VIEWING AND ROMANTIC RELATIONSHIPS - THE DOWNSIDE: A CULTIVATION EXPLORATION

Amir Hetsroni
Department of Media and Visual Arts,
Koç University, Istanbul, Turkey

ABSTRACT

This chapter reviews a series of studies that examined the connection between TV viewing and the quality of romantic relationships. The results indicate a negative pattern of associations between excessive viewing and indicators of the quality of intimate relationships and that heavy TV viewing nurtures Lodus, Pragma and Eros love styles. Romantic battles on the screen differ significantly in their reason from similar conflicts in the real world. Altogether, it seems that TV viewing is not healthy to the development of a committed long lasting relationship. The findings are analyzed mainly from a cultivation perspective, but selective exposure is also discussed as a potential explanation.

INTRODUCTION

Theoreticians and empiricists alike contend that excessive TV viewing contributes to the development of overly optimistic romantic expectations

(Segrin & Nabi, 2002) and idealistic view of marriage and coupleship (Galician, 2004; Haferkamp, 1999; Holmes, 2007; Holmes & Johnson, 2009), but what is the effect of such routine exposure to television on feelings pertaining to an already existing relationship? This is the question which I aimed to answer in a series of studies (Hetsroni, 2012a; Hetsroni, 2012b; Reizer & Hetsroni, 2014; Hetsroni & Reaizer, 2014). This article includes highlights from those studies and some aftermath insight. Readers who are interested in a closer analysis of the findings are directed to consult the original references.

My interest in the connection between TV viewing and relationship quality relies on two pillars. The first is empirical and consists of various studies which detected a pattern of *negative associations between television viewing and general mood, well-being and optimism.* Tan and Tan (1986) found that viewing of soap operas correlates with lower happiness and a sense of loneliness. Baran (1976) found that watching TV programs where sex is a leitmotif correlates with lower satisfaction with one's own sexual activities. Morgan (1984) found that lower satisfaction with life correlates with heavy TV viewing, while Bruni and Stanca (2008) found a significant association between excessive TV viewing and higher scores on a pessimism scale. Finally, Hetsroni, Sheaffer, Ben-Zion and Rosenboim (2014) found that higher exposure to television (but not to other media) correlates with pessimistic expectations about personal financial matters.

The second pillar is *cultivation theory* which proposes that since television serves as a common mode of communication and a central arena for emotional stories, heavy TV viewing may lead viewers to accept estimations and views that are represented on the screen in a disproportionate rate (compared to their presence in daily life) as an actual reflection of the world (Morgan & Shanahan, 2010; Shanahan & Morgan, 1999). The effect is an internalization of norms, beliefs and evaluations of social reality that are routinely depicted on TV rather than a copycat imitation of what occurs on the screen (Gerbner, Gross, Morgan, Signorielli & Shanahan, 2002). Cultivation can be interpreted as having occurred when a significant relationship exists between the amount of time devoted habitually to TV viewing and the adoption of views, assessments and behaviors that are often presented on the screen (Morgan & Shanahan, 2010). Cultivation proponents emphasize the cross-cultural robustness of the effect (Shanahan & Morgan, 1999) and its consistency across different demographic groups and topics (Morgan & Shanahan, 2010). They admit that size-wise a 0.1 correlation between TV viewing and reality perception (which is what a meta-analysis reveal to be an average cultivation effect) cannot be regarded as anything but a small effect, but add that television should not be expected to become a stronger socialization agent than the family or close friends and cannot shape one's life more than physical health does (Gerbner et al., 2002). Furthermore, since the effect of viewing is accumulative – over the years it may mount to a considerable influence (Shanahan & Morgan, 1999).

The cultivation effect has been traditionally divided into two types: The first refers to the positive relationship between excessive viewing and distorted perception of the world, e.g., erroneous assessment of the prevalence of romantic conflicts on certain topics in direct correspondence with these conflicts' screen presence known as a *first-order effect*. The second is the interrelatedness between excessive viewing and attitudes that directly derive from televised messages like believing that "love conquers all" as romantic drama shows, termed as a *second-order effect* (Hawkins & Pingree, 1982). Nabi and Sullivan (2001) added a *third-order cultivation effect* denoting the relationship between heavy TV viewing and behavioral intentions whose content relates directly to second-order views and attitudes. A relevant example is dating plans (to go out to a dimly-lit coffee shop and not to a fast-food joint). Romantic relationship variables usually measure second-order effects because they encompass an amalgam of attitudes toward and views of love that can be internalized from exposure to TV programming due to their ubiquitous appearance in shows. In some cases these indicators can be regarded as measures of third-order effects.

Cultivation most likely occurs because people encode information, sometimes intentionally and more often unintentionally, while they watch television. This information is retained in their long-term memory and translates into evaluations and views about the reality. This should happen more often to heavy TV viewers, since these people have encoded more TV-originated information in their memory (Shrum, Burroughs, & Rindfleisch, 2004). Indeed, there is abundant evidence that heavy TV viewing correlates with idealistic, unrealistic and sometimes even old-fashioned beliefs concerning romance that are typical of popular programming. Signorielli (1991), for instance, found a small positive relationship between TV viewing and a positive evaluation of loyalty in marriage. Holmes (2007) found a small-to-moderate positive correlation ($r = .28$) between preference for television programs with a relationship or romance theme and the romantic belief in predestined soul mates. Some dysfunctional relationship beliefs such as "mindreading is expected" and "the sexes are different" were also found to be related to TV viewing (Haferkamp, 1999; Holmes & Johnson, 2009). Segrin and Nabi (2002) found that watching programs whose main theme is romance associates with demanding ideal standards from one's partner and high marital expectations. Of course, it is possible that people who, from the start, hold idealistic unrealistic beliefs about romantic relationships bring themselves to watch TV programs that convey messages that are congruent with their beliefs, and that such self-selection co-exists alongside cultivation (Nabi, 2009). The last explanation relies on the idea of instrumental viewing or selective exposure which is a rival explanation of cultivation because it suggests that people choose to consume media that satisfies pre-existing needs (Rubin, 2002). My work does not pretend to answer which explanation is more accurate – a question that large-scale longitudinal studies have so far failed to answer (possibly because both processes co-exist), but I do aim to document the relationship between TV viewing and relationship quality assessment.

Before I present highlights from my studies I would ask to take notice of the fact that the love related message of the programming that allegedly affects the viewers is not necessarily unequivocal. Clearly, the relationship between characters in some classic hit shows like Paul and Jamie in *Mad About You*, Cory and Topanga in *Boy Meets World* and Lorelai and Luke in *Gilmore Girls* is passionate and intense and is typical of Eros love (Galician, 2004, p. 72), but recently shows began to feature less-romantic domestic arrangements (e.g., cohabitation) like Schmidt, Nick and Jess in *New Girl*. Galician (2004) insists that, even in its least romantic moments, television still connotes romanticism, but it is not out of the question that television conveys concurrently romantic and anti-romantic messages that push the viewers mainly not to be swayed by ever-lasting romance (Schreiber, 2006).

THE RELATIONSHIP QUALITY INDICATORS STUDY

This study was conducted in Israel in 2011. One-hundred and eighty eight undergraduate students (46 men and 142 women) from a public university, who were at the time of the study involved in a romantic relationship, were surveyed to assess the strength of association between routine media consumption and relationship quality. The research hypotheses were:

H1: Television viewing would contribute negatively to relationship satisfaction.

H2: Television viewing would contribute negatively to relationship commitment.

H3: Television viewing would contribute positively to the intensity of conflicts in the relationship.

H4: Compared to internet use and newspaper reading - television viewing would contribute more to the variance in relationship quality (i.e., relationship satisfaction, relationship commitment, relationship conflict).

H5: Total TV viewing time would contribute more than relationship programming viewing time to the variance in relationship quality indicators (satisfaction, commitment, conflict).

A major theoretical rationale was the negative pattern of associations between television viewing and indicators of well-being in general (Bruni & Stanca, 2008; Tan & Tan, 1986) and relationship satisfaction in particular (Holmes & Johnson, 2009). As sub-rationales served the stronger correlations of TV viewing with views and beliefs about relationships - compared to the correlation between these beliefs and consumption of other media such as newspapers (Tan & Tan, 1986) or internet (Johnson & Holmes, 2011) and the extant cultivation literature that indicates stronger ceffects for total viewing and weaker effects for exposure to specific genres (Bilandzic, & Rössler, 2004).

METHODS AND RESULTS HIGHLIGHTS

The role of dependent variables was reserved to three indicators of relationship quality: commitment, satisfaction, and conflict(s). These indicators include both the positive (satisfaction, commitment) and negative (conflict) elements of relationship quality.

Relationship satisfaction was defined as the overall subjective evaluation of the romantic relationship (Sternberg & Hojjat, 1997) and was assessed by a six-item scale developed by Rusbult, Martz and Agnew (1998). Respondents rated the fitness of statements such as “I am committed to maintaining my relationship with my partner” and “I want our relationship to last for a very long time” to their current relationship on a seven-point scale. The scale exhibited high internal consistency (α = .80). Its one factor structure was validated in a confirmatory factor analysis (RMSEA = .08, NFI = .971, CFI = .976). The idea behind this scale is that people feel satisfied, if they evaluate the relationship as meeting or exceeding a set of internal standards for a good relationship and do not perceive any other relationship as meeting those standards better (Kelley & Thibaut, 1978).

Relationship commitment was defined as the intention and desire to maintain the relationship in the long run (Rusbult & Buunk, 1993). To measure this concept we used a seven-item scale composed by Hendrick (1988). Respondents rated their agreement with sentences such as "My partner meets my needs" and "In general, I'm satisfied with my relationship" on a scale ranging from 1 (not at all) to 7 (very much). The scale exhibited high internal consistency (α = .85). Its one factor structure was validated in a confirmatory factor analysis (RMSEA = .08, NFI = .979, CFI = .986).

Conflict(s), an indicator of the downside aspects of relationships, was measured by a five-item scale adapted from Braiker and Kelley (1979). This scale measures the frequency of overt behavioral conflicts and direct communication of negative affect and overt antagonism. Respondents rate their agreement with statements such as "Our arguments often involve quite serious problems or issues" on a seven-point scale ranging from 1 (not at all) to 7 (very much). The scale exhibited adequate internal consistency (α = .79). Its one factor structure was validated in a confirmatory factor analysis (RMSEA = .01, NFI = .996, CFI = .999).

Media consumption measures included the time devoted to TV viewing on weekdays and on weekends, the time dedicated to watching certain genres, the time dedicated to Internet use (not for work purposes) and Newspaper reading. The design also included a number of *control variables:* sex, the number of romantic relationships lasting three months or more that the respondents have been involved in, and current relationship status (married or dating).

To examine the combined contribution of media consumption and control variables on relationship quality indicators and test the research hypotheses we conducted three hierarchical regression analyses. The dependent variables were relationship satisfaction, relationship commitment and relationship conflicts. The analyses are presented on Table 1 (commitment), Table 2 (conflicts) and Table 3 (satisfaction).

The hierarchical regression analysis with satisfaction as the dependent variable accounted for 8.6% of the variance, with the time dedicated to watching relationship programming accounting for most of the variance. The hierarchical regression analysis with conflicts as the dependent variable accounted for 12.5% of the variance. Watching relationship programming covered here most of the variance. The hierarchical regression analysis with commitment as the dependent variable accounted for 10.3% of the variance. Viewing relationship programming was not a significant predictor this time, but the total amount of time dedicated to television viewing accounted for most of the variance.

Table 1. Hierarchical Linear Regression Model for Relationship Commitment

Variable	Step 1					Step 2				
	Control variables					Media consumption				
	B	SE	B	*t*	*p*	B	SE	β	*t*	*p*
Sex	.001	.149	.001	.008	.994	-.011	.150	-.006	-.076	.939
Relationship status	-.246	.149	-.122	-1.659	.099	-.235	.153	-.116	-1.537	.126
Number of romantic relationships	-.077	.041	-.140	-1.894	.070	-.048	.041	-.087	-1.157	.249
Relationship programming viewing						.082	.099	.070	.829	.408
Action-adventure viewing						-.074	.099	-.063	-.745	.457
Total TV viewing						-.134	.059	-.209	-2.265	.025
Internet use						-.060	.034	-.141	-1.758	.090
Newspapers						.086	.163	.040	.525	.600
R^2			.03					.103		
$F_{(8,180)}$			1.177					2.551*		

Gender (0-male; 1-female); Relationship status (0-dating; 1-married).
$P < .05$.

To test H1, according to which television viewing contributes negatively to relationship satisfaction, we look at Table 3. Although the total amount of time dedicated to TV viewing is not significantly related to lower relationship satisfaction (β = .012, *P* > .05), the time dedicated to watching relationship programming is significantly associated with lower relationship satisfaction (β = -.211, *P* = .014). Thus, H1 is partly supported.

To test H2, according to which television viewing contributes negatively to relationship commitment, we look at Table 1. The total amount of time dedicated to TV viewing significantly associates with lower commitment to the relationship (β = -.209, *P* = .025). Thus, H2 is supported.

To test H3, according to which television viewing contributes positively to the intensity of relationship conflicts, we look at Table 2. Although the total amount of time dedicated to TV viewing is not significantly associated with more conflicts (β = .112, *P* > .05), the time dedicated to watching relationship programming does significantly associate with more relationship conflicts (β = .259, *P* = .002). Thus, H3 is partly supported.

To test H4, according to which television viewing contributes more than internet use and newspaper reading to relationship satisfaction, relationship commitment and relationship conflicts, we compare the values of TV viewing regression coefficients with those of internet use and newspaper reading. Across the board, of all the media consumption measures, only values pertaining to TV viewing account for a significant amount of variance. Therefore, H4 is supported.

Finally, to test H5, according to which total TV viewing time contributes more than the time devoted to relationship programming viewing to the variance in relationship satisfaction, relationship commitment and relationship conflicts, we compare the values of the regression coefficients pertaining to TV viewing.

Table 2. Hierarchical Linear Regression Model for Relationship Conflicts

Variable	Step 1 Control variables					Step 2 Media consumption				
	B	SE	B	*t*	*p*	B	SE	β	*t*	*p*
Sex	.106	.196	.041	.541	.589	.021	.192	.008	.111	.912
Relationship status	-.120	.195	-.046	-.614	.540	.024	.196	.009	.120	.904
Number of romantic relationships	-.060	.053	-.083	-1.118	.265	-.053	.053	-.074	-1.008	.315
Relationship programming viewing						.394	.127	.259	3.111	.002
Action-adventure viewing						.036	.127	.024	.282	.778
Total TV viewing						.093	.076	.112	1.225	.222
Internet use						.033	.044	.059	.750	.454
Newspapers						.060	.209	.022	.287	.774
R^2			.011					.125		
$F_{(8,180)}$			.688					3.160**		

Gender (0-male; 1-female); Relationship status (0-dating; 1-married).
** $P < .005$.

Table 3. Hierarchical Linear Regression Model for Relationship Satisfaction

Variable	Step 1 Control variables					Step 2 Media consumption				
	B	SE	B	*t*	*p*	B	SE	β	*t*	*p*
Sex	.182	.138	.099	1.318	.189	.198	.139	.107	1.422	.157
Relationship status	-.096	.137	-.052	-.697	.487	-.140	.142	-.075	-.986	.325
Number of romantic relationships	.063	.038	.125	1.688	.093	.055	.038	.108	1.428	.155
Relationship programming viewing						-.228	.092	-.211	-2.481	.014
Action-adventure viewing						-.113	.092	-.106	-1.234	.219
Total TV viewing						.007	.055	.012	.125	.901
Internet use						-.014	.032	-.034	-.426	.671
Newspapers						-.061	.151	-.031	-.402	.688
R^2			.023					.086		
$F_{(8,180)}$			1.419					2.080*		

Gender (0-male; 1-female); Relationship status (0-dating; 1-married).
$^*P < .05$.

In commitment, total viewing time is a significant regression coefficient (β = -.209, *P* = .025), whereas viewing relationship programming is not significant (β = .070, *P* > .05). In satisfaction and conflicts, on the other hand, viewing relationship programming is a significant regression coefficient (β = -.211, *P* = .014 and β = .259, *P* = .002 respectively), while total viewing time is not significant (β = .012, *P* > .05 and β = .112, *P* > .05 respectively). In light of this mixed pattern, H5 is not supported.

Discussion Highlights

After controlling for factors such as sex, relationship status and the number of previous romantic engagements, TV viewing remains significantly associated with positive and negative indicators of relationship quality.

From a cultivation perspective, while our results show that television viewing is still a stronger predictor of relationship quality than exposure to newspapers or to new media (*Cf.* Johnson & Holmes, 2011), possibly due to newspapers' tendency to concentrate on politics and the minor presence of romantic themes in the most popular internet portals (Orr, 2004), they do not entirely support the theory's presumption that general TV viewing is always a better predictor than genre specific viewing (See Bilandzic & Rössler, 2004 for a review). While we found that routine exposure to TV shows that are totally unrelated to romance (e.g., *action-adventure programming*) is not statistically predicting any indicator of relationship quality, our findings nonetheless suggest that both general viewing and genre specific viewing should be considered as potentially affecting factors, since total viewing time performs better in predicting commitment to the relationship while the amount of time devoted to watching relationship programming is a better predictor of satisfaction from the relationship and the intensity of conflicts within it. Commitment is an unusual measure among the three relationship indicators because it is represents a behavioral inclination to maintain a romantic relationship (Schoebi, Karney & Bradbury, 2012). Thus, it is an assessment of a third-order cultivation effect (Nabi & Sullivan, 2001). This study can be classified as one of the very few to confirm this type of effect. Indeed, the fact that general viewing is linked to commitment (but not to satisfaction and conflicts) supports the special position of commitment as the more behavioral measure of relationship quality assessment.

THE LOVE STYLE STUDY

The relationship between having certain attitudes towards love and different levels of TV viewing is another association that can be explained by *cultivation theory*. It is also another area where the contribution of TV viewing to the development of a negative attitude toward romance is noticed. As we shall see, TV viewing nurtures a sometimes pragmatic-suspicious sometimes uncommitted conception of intimate relationship.

THE CONCEPT OF LOVE STYLE

In his book "The Colors of Love" sociologist John Alan Lee (1973) suggested a *love style taxonomy* which was applied to a *Love Attitude Scale* that has been validated in various cultures (Hendrick & Hendrick, 1986). Lee's taxonomy consists of six love styles: Eros, Ludus, Storge, Pragma, Mania and Agape. These styles translate into different types of lovers that are described by Hendrick (2004) as follows: the *Eros* lover is passionate. His love is intense and he wants to be involved with his partner at all levels. He is self-confident and trusting of his partner. Culturally, Eros love is known as romantic. The *Ludus* lover, in contrast, is not interested in intensity but rather experiences love as a game to be played, not necessarily with serious intentions. Ludic love is often played with several partners at a time, so that different people may be enjoyed for different qualities in different activities. This love style has a lot to do with what is culturally known as flirting. The *Storge* lover uses a strong sense of friendship to establish partnership that is based on similarity and completion in terms of attitudes and values. This similarity is more important to the Storge lover than physical appearance or sexual satisfaction because he prefers long-term commitment to short-term excitement. Culturally, this love style is known as soul mating. The *Pragma* lover, as the name implies, is practical. The pragmatic lover is looking for a suitable partner with whom a rewarding life can be built. To achieve that, he has a shopping list of qualities sought in a partner. Love here is treated as a kind of transaction in which physical attraction is only part (and not always the most crucial part) of the deal. The *Mania* lover is characterized by emotional highs and lows, possessiveness, dependence, jealousy and insecurity. Another component of Mania love may be physical symptoms such as difficulty in eating or sleeping. The *Agape* lover is characterized by altruism to the extent that the partner's welfare is more important than what he himself gets from the relationship. Commonly known as compassionate love, the Agape lover attributes little importance to sensuality and sexuality and cares more about spiritual qualities.

Most people are dominated by one or two styles (Hendrick & Hendrick, 1992) that determine a lot of what happens in their love lives. For instance, Eros predicts our satisfaction with long term romantic relationships (Meeks, Hendrick & Hendrick, 1998), whereas Ludus predicts lack of satisfaction with such relationships (Contreras, Hendrick & Hendrick, 1996). Love style are also connected (see Hendrick & Hendrick, 1992, for review) with personality traits (self-esteem, for instance, is positively connected with Eros and negatively associated with Mania), ethnicity and culture (westerners are higher in Pragma compared to people from traditional cultures), income (in comparison with poor people - rich people are higher in Pragma), sex (females are higher in Pragma and Mania and are lower in Agape and Ludus) and age (older people are less storgic and younger people are less pragmatic). People who are currently involved in a committed partnership are higher in Storge and Agape, whereas people who are not at the moment part of such partnership are higher in Ludus (Montgomery & Sorell, 1997). Most of the studies about love styles use college student samples, because young people are likely to find love an engaging topic on which they are ready to share their opinion (Hendrick & Hendrick, 2002). Our study was not an exception. The aims were to find potential connections between TV viewing and certain love styles and to try and explain these connections. The rationale was that since the different types of programs promote different love styles – different viewing patterns should emerge and can be interpreted either as a second order cultivation effect or as an outcome of selective exposure. Specifically, we expected to find a positive correlation between watching love related programming and endorsing Eros love style because a previous study detected a significant positive association between watching romantic comedies, soap operas, daytime talk shows, and reality programs about romance and higher levels of Eros (Segrin & Nabi, 2002) and left other connections between TV viewing and love styles open for exploration.

METHODS AND RESULTS HIGHLIGHTS

This study was conducted in Israel in March-April 2010 and surveyed 338 unmarried undergraduate students attending a public university for their love style, TV viewing habits, demographic characteristics, and love status (current involvement in a serious romantic relationship and marital intentions). The questionnaire included measures of TV viewing habits, control variables and the *Love Attitude Scale* which was composed by Hendrick and Hendrick (1986) as a measure of Lee's *six love styles*. The 42 items, seven items per love style in each sub-scale, refer to a specific love relationship at present (when the respondent is currently in a relationship), or to a past relationship (when the respondent is not currently in a relationship but was in a relationship in the past), or to the conception of a relationship (when the respondent was never part of a romantic relationship), and offer two versions – one worded for males and one worded for females. The respondents are asked to mark their agreement with statements on a scale that ranges from 1 ("totally disagree") to 5 ("totally agree"). Here are sample items:

> *Eros:* "My partner and I have the right physical 'chemistry' between us."
>
> *Ludus*: "I believe that what my partner doesn't know about me won't hurt him/her."
>
> *Storge*: "Our love is the best kind because it grew out of a long relationship."
>
> *Pragma:* "A main consideration in choosing my partner is how s/he would reflect on my family."
>
> *Mania*: "When my partner doesn't pay attention to me, I feel sick all over."
>
> *Agape*: "I would rather suffer myself than let my partner suffer."

The sub-scales reliability coefficient values were as follows: Eros (α = .71), Ludus (α = .67), Pragma (α = .72), Mania (α = .70) and Agape (α = .79). These values denote an adequate level of reliability. In contrast, the alpha Cronbach coefficient pertaining to Storge (α = .55) was less than sufficient, but it crossed the 0.6 threshold (α = .64), after one of the items ("It is hard for me to say exactly when our friendship turned into love") had been omitted from the scale. A confirmatory factor analysis performed on the revised instrument (41 statements) tested the orthogonal six-factor model, as stipulated by Hendrick and Hendrick (1990), yielded a goodness of fit index (GFI) that is equal to .90 and root mean square error of approximation (RMSEA) that is equal to .04. The items' factor loadings were all above the .40 threshold and were all statistically significant ($p < .05$). These figures attest that the multidimensionality of the scale fits into the theoretical six-factorial structure.

For initial inspection of the connections between TV viewing and various love styles Pearson correlations between general viewing, news viewing, and viewing of love-related programming and each of the six love styles were computed. They are presented on Table 4.

Only three love styles correlated significantly with TV viewing: Ludus (correlated positively with general viewing and with watching news programs); Pragma (correlated positively with viewing of love related programming but negatively with news programs viewing); and Eros (correlated positively with general viewing and with viewing of love related programming). The last correlation supports our hypothesis. In the second stage of the analysis, the love styles that significantly correlated with TV viewing – Eros, Ludus and Pragma – were used as dependent variables in three hierarchical linear regression procedures. Entered in three steps, the predictors were sex, ethnicity, and income (block I – demographic characteristics), current involvement in a serious romantic relationship and anticipation of getting married in the foreseeable future (block II – love status), and general television viewing, news programming viewing, and viewing of love related programming (block III – TV viewing). The results of the regression procedures appear on Table 5.

The regressions indicate that under multiple controls general TV viewing predicts a higher level of Ludus (β = .136, P = .04). News programming viewing predicts higher levels of Ludus (β = .189, P = .002) and Pragma (β = .222, $P < .001$). Viewing of love related programming predicts a higher level of Eros (β = .127, P = .03) and a lower level of Pragma (β = -.181, P = .003). The association of Eros and general viewing is no longer significant in the multivariate model.

Table 4. Bivariate Pearson Correlations between TV viewing and Love Styles (*N = 338*)

	Eros	Ludus	Storage	Pragma	Mania	Aga
TV Viewing - General (Daily) (M=145min. ; SD=49min.)	.12 *	.18 **	.06	.00	.04	.07
TV Viewing – News (Daily) (M=40min. ; SD=18min.)	-.01	.29 **	.04	.14 *	.03	.11
TV Viewing – Love Related Programming (Daily) (M=75min. ; SD=29min.)	.12 *	.02	.00	-.15 *	.04	.07

* $P < .05$, ** $P < .01$.

DISCUSSION HIGHLIGHTS

The findings are generally consistent with a small effect from exposure to television programming to differences in love styles. A cultivation explanation of the significant correlations is that perennial flirting (Ludus), seeking socially worthwhile relationships (Pragma), and passionate romance (Eros) are tendencies that can be internalized - to some extent - from watching television.

The association of Eros with viewing of love related shows replicates the results of a previous study (Segrin & Nabi, 2002) and comes to terms with the prediction of cultivation theory that watching programs such as soap operas and family dramas involving romantic themes would create a tendency to have romantic Eros characteristics because of the romantic nature of these genres (Galician, 2004; Schreiber, 2006). This can also explicate why Pragma, the least romantic of all love styles, is negatively correlated with watching them. If we also take into account the positive correlation between Pragma and viewing of news programming, the overall picture, regarding Pragma, is of a practical approach: People with Pragma tendencies choose to watch news programming that contain valuable information about the surroundings, are interested in socially worthwhile relationships, and avoid romantic programming. The negative correlation of Eros with news programming is, therefore, placing Eros as the opposite of Pragma, when it comes to TV viewing habits. The significant (and highest across the board) correlation of news programming with Ludus appears inconsistent with cultivation theory. Literal facts about the world that are commonly featured in the news seem to be unrelated to the key definitional characteristics of Ludus. Watching programming that does not feature any flirting element should not cultivate Ludus tendencies. However, perhaps, the suspicion and the lack of commitment embedded in many news items e.g., reports about political intrigues may appeal to people with Ludus tendencies, who are essentially non-committed to a single partner (Hendrick, 2004). We did not find any meaningful correlations between the amount of time devoted to TV viewing and the level of Storge, Agape, and Mania because (if we follow the logic of cultivation theory) these love styles are very rarely featured in popular programming.

Obviously, because it is cross-sectional study the data can also be interpreted in reverse from cultivation theory. In other words, it is not out of the question that people who have a tendency for a love style(s) choose to expose themselves intentionally to programming in accord with that style. For example, Eros lovers may choose to watch love related programming because this programming, which is full of descriptions that are in line with their character (Segrin & Nabi, 2002), can satisfy their various social and psychological needs and interests (Rubin, 2002). Although selective exposure is an alternative interpretation to cultivation theory, it is also possible that both processes operate simultaneously.

Table 5. Hierarchical Linear Regression for Eras, Ludus and Progma Love Styles ($N = 338$)

Block		Eros Love Style B	Eros Love Style β	Eros Love Style Partial r	Eros Love Style t	Ludus Love Style B	Ludus Love Style β	Ludus Love Style Partial r	Ludus Love Style t
Demographic measures									
	Sex (0-male ; 1-female)	-.178	-.136	-.144	-2.248 *	-.361	-.249	.268	-4.44
	Ethnicity (0-Arab ; 1-Jew)	.310	.058	.062	.950	.115	.010	.011	.17
	Income	-.030	-.051	-.054	-.837	-.004	-.001	.001	-.0
Incremental adjusted R2		.018 *				.083 ****			
Love life									
	Romantic relationship (0-no ; 1-yes)	.333	.273	.278	4.462***	-.118	-.065	-.072	-1.1
	Foreseeable future marriage (0-no ; 1-yes)	.269	.156	.164	2.552 *	-.446	-.193	-.208	-3.41
Incremental adjusted R2		.103****				.055 ***			
Television viewing									
	General	.010	.054	.052	.801	.054	.136	.140	2.26
	News	.004	.005	.005	.073	.070	.189	.189	3.08
	Love related programming	.056	.127	.136	2.109 *	-.030	-.086	-.086	-1.38
Incremental adjusted R^2		.016 *				.071 ***			
Adjusted model R^2		.136 *****				.209 ****			
Intercept		4.039			9.555****	2.039			5.352

* $P < 05$, ** $P < .01$, *** $P < 005$, **** $P < .001$.

"WHAT THE FIGHTING IS ALL ABOUT" STUDY

This study compared the topics for fighting in TV programming with what real-world couples mention as reasons for domestic battles. The theoretical framework here was reality construction which is to a great extent part of the cultivation hypothesis. Cultivation researchers content that television features a distorted by yet captivating and believable picture of the world. This is exactly the idea of reality construction: Because the media systematically refrains from duplicating reality one by one and (for dramatic and commercial purposes) prefers to highlight certain elements in the picture and suppress others, we should not expect to find a close match between the TV representation of the world and actual circumstances (Weimann, 2000).

In real life, couples fight mainly about money (Amato & Rogers, 1997; Dew, Britt & Huston, 2012) and communication problems (Papp Cummings & Goeke-Morey, 2009). Sex is rated lower in importance, particularly in comparison with jealousy (Papp, Goeke-Morey & Cumming, 2013). The impact of extended family members is not often mentioned as a reason for fighting, but disagreements about relationship with friends and household chores are stated as a "hot topic" by both husbands and wives (Papp et al., 2009; Williamson et al., 2013). While men and women do differ in the importance they ascribe to sex (Williamson, Hanna, Lavner, Bradbury & Karney, 2013), there are no major gender differences about the detrimental effect of miscommunication. The same goes with disputes about raising and educating the children (Papp et al., 2009). In general, there are more similarities than differences between the genders in what they see as "a good reason for fighting" perhaps because in a long term relationship that becomes a storgic friendship both men and women develop rather similar needs (Finchan & Beach, 1999).

On television, fighting between romantic partners is strongly shaped by dramatic needs (Silverstone, 1994). To break-up over house chorus disagreement might be a realistic scene but not a very strong drama. Of course, we should not forget the indirect didactic power of popular programming (e.g., from watching break-up scenes on television we may learn techniques of parting ways) and the cultivation effect, which may make us believe that some of the reasons for fighting are more common and normative than they actually are (Gerbner et al., 2002). In addition, as always, mapping daily life situations on the small screen may portray social trends and indicate what is acceptable in society (Greenberg, 1980, p. XII). In the past, popular programming (e.g., Ozzie & Harriet, Father Knows Best) cherished family values in an all-White trouble free suburbia and framed romantic conflicts as something ephemeral (Lichter, Lichter & Rothman, 1994) and limited to minor solvable topics such as food choices, dress code, etc. Preferring non-objectionable shows was an ample tactic in a broadcast environment predominated by a small number of networks operating in a rather conservative society. In recent decades, the sexual revolution and along with it a more permissive approach to relationship fidelity and an acceptance of non-traditional families found an expression on television as well. Popular programming is no longer the Happyland it used to be in the 1950s and 1960s, but the majority of conflicts on television still have a happy ending which transmits the serroneous myth that a conflict is good for the relationship (Buslig & Ocaña, 2007). The lack of previous mapping of the distribution of conflict topics on television directed us not to pose exact hypotheses, but ask a research question:

RQ: Is there a difference in the distribution of conflict topics between TV programming and real life ranking ?

Methods and Results Highlights

To answer the question we conducted a content analysis of TV programming and ran a survey among young adults (who are engaged in a serious romantic relationship). The respondents' ranking of conflict topics was juxtaposed with the findings of the content analysis of romantic conflicts in TV programs.

Survey

The survey was conducted in Israel in May 2011. One hundred and eighty eight undergraduate students (46 men and 142 women) from a public university completed the questionnaire that asked (based on Papp et al., 2009):

> There are many topics about which romantic partners, married or not married, fight. We want to know to what extent you and your partner fight about some topics. Please rank each topic from 0 to 100 - when 0 means that you do not fight at all over this topic and 100 designates that you fight a lot about the topic.

The list included the following topics: money, miscommunication, family relatives, sex, religion, leisure, friends, alcohol, drugs, children, household chores, jealousy and other issues.

Only students who confirmed that they were involved in a romantic relationship (lasting three months or longer), when the survey took place, were allowed to take part in the study. Forty-five respondents (24% of the sample) were already married, when the study took place.

Content Analysis

To create the TV conflicts sample, we picked up all the fictional programs in which a romantic narrative is central to the plot that were broadcast in 2011 on one of Israel's two most highly watched terrestrial channels (Channel 2, Channel 10) and in the two most highly watched cable/satellite channels (Hot3, Yes-Stars). Altogether, these four channels attract more than 70% of the viewers during the prime-time hours on a typical weekday (Eurodata TV, 2011). The following shows supplied conflicts to the sample (in parenthesis – whether the show was original, import, or locally cloned version of a foreign format): *Two and a Half Men* (import), *Bold and Beautiful* (import), *Traffic-Light* (local), *Life is only Part of it* (local), *Golden Girls* (local version of a foreign format), *Love and Marriage* (local version of a foreign format), Three (local), Grey's Anatomy (import), *Scrubs* (import), *How I Met your Mother* (import), *Big Bang Theory* (import), *Brothers and Sisters* (import), *2.3 kids on Average* (local). From each show three episodes were randomly selected. All in all, the analysis was based on 27 hours of programming.

Two coders (graduate students with vast experience in coding of TV content) were instructed to pay notice to conflicts between romantic partners and identify who initiated the conflict (the man or the woman) and list - from among the aforementioned topics - the most salient topic in the fight. The coders were trained in tandem for seven hours and on an individual basis for additional three hours. Each coder worked alone and each episode was coded by two different coders. To measure agreement between coders and inter-coder reliability, Cohen's Kappa statistic was computed. Its value ($\kappa = 0.82$) is impressive given the complexity of the content domain. Table 6 lists the share of each topic and its internal ranking in two databases: content analysis of TV programming and survey responses.

Table 6. Conflict Topic in Fictional TV Programming and in the Survey

Topic	Conflict topic in fictional television programming (N = 110)		Conflict topic in the survey (N = 188)	
	Share	Ranking	Share	Ranking
Children	16.5%	1	2.3%	11
Sex	15.5%	2	5.8%	10
Miscommunication	14.5%	3	14.4%	2
Jealousy	13.6%	4	9.7%	4
(Extended) Family Members	12.7%	5	10.2%	3
Friends	7.3%	6	6.9%	8
Other	6.4%	7	0.8%	12
Housework chores	5.4%	8	16.7%	1
Leisure	4.5%	9	9.5%	5
Money	3.6%	10	8.5%	6-7
Religion	0.0%	11-12	8.5%	6-7

The two databases feature a small number of similarities and several dissimilarities. In both of them, the most frequently mentioned topics hover around the 15% threshold. In both of them, miscommunication (ranked third in the programming and second in the survey) and jealousy (ranked fourth in the programming and also in the survey) are among the more often mentioned topics. However, children and sex, which are ranked first and second in the programming, where they enjoy respective shares of 16. 5% and 15.5%, are ranked only eleventh and tenth in the survey where their respective shares are merely 2.3% and 5.8%. Fights over money, on the other hand, are mentioned more often by the respondents (8.5%) than they appear in the programming (3.6%). The lack of strong similarity between the two databases is confirmed by the negligible value of the correlation between the lists ($rs = -.075$, $P > .05$). Thus, the answer to the research question is that the relative prevalence of most of the topics of romantic conflicts presented in popular programming is different from their relative prevalence in real life.

DISCUSSION HIGHLIGHTS

The findings indicate minor similarities and major differences between the topics of romantic conflicts presented on television and the reasons for conflict that are reported by young adults. Miscommunication ranked nearly equally high in television and in the survey (as it did in previous surveys – see Papp et al., 2009), possibly because it has a crucial impact on daily life and yet also has a notable dramatic value. However, children, sex, and money - pointed in previous studies as highly frequent reasons for fighting among couples (Dew & Dakin, 2011; Papp et al., 2009) – were found to be negligible in our survey (but not so much in the programming) while household chores, which were found to be the most prominent topic of conflict in the survey (in line with the results of previous studies) was hardly notable in the programming.

The pattern of survey findings may have to do with the demographic makeup of our sample. Twenty-something respondents can be too young to have children. They may also be too religious (56% defined themselves as observant or at least traditional) to allow themselves to admit publicly (even anonymously) that they argue about money and sex. The sizeable presence of religion as a source of conflict in our survey, far beyond what was found previously in western societies (Oggins, 2003), gives further credence to the last argument. As for the programming, the prominence of sex, jealousy and children as sources of conflict is in line with their centrality in TV drama (Marc, 1996) and their visual appeal that can attract viewers.

The minimal resemblance in ranking of romantic conflicts topics between the programming content and the viewers' ratings is in line with the idea that television features a distorted picture of reality (Weimann, 2000). Romantic reality, as reconstructed on the home screen, takes place in a land where partners mainly fight about sex and money and sometimes quarrel about their children, whereas in the real world they are more concerned with housework labor (which is less dramatically appealing and is therefore missing from the screen).

GENERAL CONCLUSION

The combined picture that emerges from the three studies is somewhat worrying. We were unable to point at even one positive association between TV viewing and relationship quality, but we did find plenty of evidence that

TV viewing is detrimental to the strength of intimate relationships, that it encourages the development of non-lasting or non-committed love style and presents a very distorted picture of domestic conflicts where the fighting is fundamental and full of passionate drama while ignoring mundane conflicts that can be easily resolved.

These results are in line with the assumption that unrealistic beliefs, expectations and standards based on distorted assumptions impair the subjective assessment of intimate relationships quality and foster disappointment from the dyadic connection (Sharp & Ganong, 2000), mainly because people feel disappointed when their expectations are not met. TV portrayal of romantic partners as ideal types jeopardizes the chances of heavy viewers to be satisfied with their current relationship (Galician, 2004), since the ideal standards for romance on the small screen operate as accessible knowledge structures (Shrum et al., 2004). It is the presentation of the 'ideal' as 'typical' that arguably makes media images influential and potentially harmful for heavy media consumers, who may feel as if their own relationships do not measure up to those observed media ideals. In other words, the ideal portrayal of a happy couple cuddling in front of the tube and enjoying and romantic show lacks scientific basis.

From a cultivation perspective, our results show that television viewing is still a stronger predictor of relationship quality than exposure to newspapers or to new media (*Cf.* Johnson & Holmes, 2011), possibly due to newspapers' tendency to concentrate on politics and the minor presence of romantic themes in the most popular internet portals (Orr, 2004), but we have contradicting findings regarding the differences in impact between general viewing and genre-specific exposure: The first was a stronger predictor of relationship quality and the second was a stronger predictor of love styles. Cultivation proponents suggest that the effect genre-specific exposure is actually better conceived as an outcome of selective exposure (Morgan & Shanahan, 2010). Indeed, we are unable to completely rule out selective exposure as an alternative interpretation of the findings, but the direction of the influence is possibly not crucial because our results adjoin a growing body of literature that unequivocally relates TV viewing to pessimism (Hetsroni et al., 2014; Bruni & Stanca, 2008), lack of happiness (Tan & Tan, 1986) and low level of satisfaction from life (Baran, 1976; Morgan, 1984). The correlations we found, while not particularly high in magnitude, are still in some cases stronger than what the cultivation research usually typically finds.

Obviously, in studies such as ours the location may mean a lot. Had the relationship conflicts research taken place in America for instance and not in

Israel, the respondents might have ranked money as a more important conflict topic in their life (and then the gap between what the programming shows about domestic conflicts and what people think of them might have been smaller). This is a recommendation to conduct replications in various countries to cross-culturally validate the findings.

REFERENCES

Amato, P. R. and Rogers, S. J. (1997). A longitudinal study of marital problems and subsequent divorce. *Journal of Marriage and the Family*, 59, 612-624.

Baran, S.J. (1976). How TV and film portrayals affect sexual satisfaction in college students. *Journalism Quarterly*, 53, 468-473.

Bilandzic, H., and Rössler, P. (2004). Life according to television: Implications of genre-specific cultivation effects: The gratification/cultivation model. *Communications: The European Journal of Communication Research*, 29(3), 295-326.

Braiker, H. B., and Kelley, H. (1979). Conflict in the development of close relationships. In: R. L. Burgess and T. L. Huston. (Eds.) *Social exchange in developing relationships* (pp. 135-168). NY: Academic Press.

Bruni, L., and Stanca, L. (2008). Watching alone: Relational goods, television and happiness, *Journal of Economic Behavior and Organization* 65, 506-528.

Buslig, A. L. S., and Ocaña, A. M. (2007). Myths of romantic conflict in the television situation comedy. In: M-L. Galician and D. L. Merskin (Eds.), *Critical thinking about sex, love and romance in the mass media* (pp. 203-213). Mahwah, NJ: Lawrence Erlbaum.

Contreras, R., Hendrick, S. S., and Hendrick, C. (1996). Perpectives on marital love and satisfaction in Mexican-American and Anglo-American couples. *Journal of Counseling and Development*, 74, 408-415.

Dew, J., Britt, S. and Huston, S. (2012). Examining the relationship between financial issues and divorce. Family Relations: An Interdisciplinary. *Journal of Applied Family Studies*, 61, 615-628.

Dew, J. P, and Dakin, J. (2011). Financial disagreements and marital conflict tactics. *Journal of Financial Therapy*, 2, 23-42.

Eurodata TV. (2011). One *television year in the world*. Paris, France: Mediametrie.

Fincham, F. D. and Beach, S. R. H. (1999). Conflict in marriage: Implications for working with couples. *Annual Review of Psychology*, 50, 47-77.

Galician, M-L. (2004) *Sex, love and romance in the mass media: Analysis and criticism of unrealistic portrayals and their influence*. Mahwah, NJ: Lawrence Erlbaum.

Gerbner, G., Gross, L., Morgan, M., Signorielli, N., and Shanahan, J. (2002). Growing up with television: Cultivation processes. In: J. Bryant and D. Zillman (Eds.), *Media effects: Advances in theory and research* (pp. 43-67). Mahwah, NJ: Lawrence Erlbaum.

Greenberg, B. S. (1980). *Life on television: Content analysis of US TV drama*. Norwood, NJ: Ablex.

Haferkamp, C. J. (1999). Beliefs about relationships in relation to television viewing, soap opera viewing, and self-monitoring. *Current Psychology*, 18, 193-204.

Hawkins, R. P., and Pingree, S. (1982). Television's influence on constructions of social reality. In: D. Pearl, L. Bouthilet, and J. Lazar (Eds.), *Television and behavior: Ten years of scientific progress and implications for the eighties* (Vol. 2, pp. 224-247). Washington, DC: Government Printing Office.

Hendrick, C., and Hendrick, S. S. (1986). A theory and method of love. *Journal of Personality and Social Psychology*, 50, 392-402.

Hendrick, C., and Hendrick, S. S. (1990). A relationship-specific version of the Love Attitudes Scale. *Journal of Social Behavior and Personality*, 5, 239-254.

Hendrick, S. S. (1988). A generic measure of relationship satisfaction. *Journal of Marriage and the Family*, 50(1), 93-98.

Hendrick, S. S. (2004). Close relationships research: A resource for couple and family therapists. *Journal of Marital and Family Therapy*, 30(1), 13-27.

Hendrick, S. S., and Hendrick, C. (1992). *Romantic love*. Newbury Park, CA: Sage.

Hendrick, S. S., and Hendrick, C. (2002). Linking romantic love and sex: Development of the Perceptions of Love and Sex Scale. *Journal of Social and Personal Relationships*, 19, 361-378.

Hetsroni, A. (2012a). A cultivation perspective on love styles. *American Journal of Media Psychology*, 5, 53-68.

Hetsroni, A. (2012b). Is television viewing associated with love styles? A cultivation study. *Psychological Report*s, 110(1), 35-50.

Hetsroni, A., and Reizer, A. (2014). What's the fighting all about? A comparison of reasons for romantic conflicts on television programs and in the real world. In: M. Moshe (Ed.). *The emotions industry*. Hauppauge, NY: Nova Science Publishers.

Hetsroni, A., Sheaffer, Z., Rosenboim, M., and Ben-Zion, U. (2014). Economic expectations, optimistic bias and exposure to different television formats during economic recession - A cultivation study. *Communication Research*, 41(2), 180-207.

Holmes, B. M. (2007). In search of my "one-and-only": Romance-oriented media and beliefs in romantic relationship destiny. *Electronic Journal of Communication*, 17, 3-4.

Holmes, B. M., and Johnson, K. (2009). Where fantasy meets reality: Media exposure, relationship beliefs and standards, and the moderating effect of a current relationship. In: E. P. Lamont (Ed.), *Social psychology: New research* (pp. 117-134). Hauppauge, NY: Nova Science.

Johnson, K. R., and Holmes, B. M. (2011). Media and romantic relationships: An emerging research area. In: M. Paludi (Ed.), *The psychology of love* (pp. 89-105). Santa Barbara, CA: Praeger.

Kelley, H. H. (1983). Love and commitment. In: H. H. Kelley, E. Berscheid, A. Christensen, J. H. Harvey, T. L. Huston, G. Levinger, E. McClintock, L. A. Peplau, and D. R. Peterson (Eds.), *Close relationships* (pp. 265-314). NY: Freeman.

Kelley, H. H., and Thibaut, J. W. (1978). *Interpersonal relations: A theory of interdependence*. New York: Wiley.

Lee, J. A. (1973). *The colors of love: An exploration on the ways of loving*. Don Mills, Ontario, Canada: New Press.

Lichter, S. R., Lichter, L. S. and Rothman, S. (1994). *Prime time: How TV portrays American culture*. Washington, DC: Regnery.

Marc, D. (1996). *Demographic vistas: Television in American culture*. Philadelphia, PA: University of Pennsylvania Press.

Meeks, B. S., Hendrick, S. S., and Hendrick, C. (1998). Communication, love, and relationship satifaction. *Journal of Social and Personal Relationships*, 15, 137-142.

Morgan, M. (1984). Heavy television viewing and perceived quality of life. *Journalism Quarterly*, 61, 499-504.

Morgan, M., and Shanahan, J. (2010). The state of cultivation. *Journal of Broadcasting and Electronic Media*, 54, 337-355.

Montgomery, M. J., and Sorell, G. T. (1997). Differences in love attitudes across family stages. *Family Relations*, 46, 55-61.

Nabi, R. L. (2009). Emotion and media effects, In: R. L. Nabi and M. N. Oliver (Eds.), *The Sage handbook of media processes and effects* (pp. 131-143). Thousand Oaks, CA: Sage.

Nabi, R. L., and Sullivan, J. L. (2001). Does television viewing relate to engagement in protective action against crime? A cultivation analysis from a theory of reasoned action perspective. *Communication Research*, 28, 802-825.

Oggins, J. (2003). Topics of marital disagreement among African-American and Euro-American newlyweds. *Psychological Reports*, 92, 419-425.

Orr, A. (2004). *Meeting, mating (... and cheating): Sex, love and the new world of online dating*. Upper Saddle River, NJ: Reuters.

Papp, L. M., Cummings, E. M. and Goeke-Morey, M. C. (2009). For richer, for poorer: Money as a topic of marital conflict in the home. *Family Relations*, 58, 91-103.

Papp, L. M., Goeke-Morey, M. C. and Cummings, E. M., (2013). Let's talk about sex: A diary investigation of couples' intimacy conflicts in the home. *Couple and Family Psychology: Research and Practice*, 2, 60-72.

Reizer, A., and Hetsroni, A. (2014). Media exposure and romantic relationship quality: A slippery Slope? *Psychological Reports*, 114(1), 231-249.

Rubin, A. M. (2002) The uses-and-gratifications perspective of media effects. In: J. Bryant and D. Zillman (Eds.), *Media effects: Advances in theory and research*. (pp. 525-548). Mahwah, NJ: Lawrence Erlbaum.

Rusbult, C. E., and Buunk, B. P. (1993). Commitment processes in close relationships: An interdependence analysis. *Journal of Social and Personal Relationships*, 10, 175-204.

Rusbult , C. E., Martz , J. M., and Agnew C. R. (1998). The investment model scale: Measuring commitment level, satisfaction level, quality of alternatives, and investment size. *Personal Relationships*, 5, 357-391.

Schoebi D., Karney, B. R., and Bradbury T. N., (2012). Stability and change in the first 10 years of marriage: Does commitment confer benefits beyond the effects of satisfaction? *Journal of Personality and Social Psychology*, 102, 729-742.

Schreiber, M. (2006). *"You don't want to be in love... You want to be in love in a movie": Romance and postfeminism in contemporary film and television.* Unpublished doctoral dissertation, University of California – Los Angeles (UCLA), Los Angeles, CA.

Segrin, C., and Nabi, R. L. (2002). Does television viewing cultivate unrealistic expectations about marriage? *Journal of Communication*, 52, 247-263.

Shanahan, J., and Morgan, M. (1999). *Television and its viewers: Cultivation theory and research*. Cambridge, UK: Cambridge University Press.

Sharp, E. A., and Ganong, L. H. (2000). Raising awareness about marital expectations: Are unrealistic beliefs changed by integrative teaching? *Family Relations*, 49, 71-76.

Shrum, L. J., Burroughs, J. E., and Rindfleisch, A. (2004). A process model of consumer cultivation: the role of TV is a function of type of judgment, In: L. J. Shrum (Ed.), *The psychology of entertainment: blurring the lines between entertainment and persuasion* (pp. 177-192). Mahwah, NJ: Lawrence Erlbaum.

Signorelli, N. (1991). Adolescents and ambivalence toward marriage: A cultivation analysis. *Youth and Society*, 23, 121-149.

Silverstone, R. (1994). *Television and everyday life*. London: Routledge.

Sternberg, R. J., and Hojjat M. (1997). *Satisfaction in close relationships*. New York, NY: The Guilford Press.

Tan, A., and Tan, G. (1986). Television use and mental health. *Journalism and Mass Communication Quarterly*, 63, 106-113.

Weimann, G. (2000). *Communicating unreality: modern media and the reconstruction of reality*. Thousand Oaks, CA: Sage.

Williamson, H. C., Hanna, M. A., Lavner, J. A., Bradbury, T. N. and Karney, B. R. (2013). Discussion topic and observed behavior in Couples' problem-solving conversations: Do problem severity and topic choice matter? *Journal of Family Psychology*, 27, 330-335.

LIST OF CONTRIBUTORS

Sue Aran-Ramspott holds a PhD in Audiovisual Communication from Ramon Llull University (URL). She has taught masters courses and specialized seminars at the Blanquerna School of Communications and International Relations (URL) since its beginnings (1994), where she has served as the Director of the Media Studies Department and as the Coordinator of the Doctoral Program. She won the First Prize in Research in Audiovisual Communication from the Catalan Audiovisual Council (CAC) in 2009. She has participated in different research projects, and currently she is the leading researcher in the Research Group on Television Representations and Social Imagery (URL, UPF, UB, UPC, Barcelona). Latest publications include entries in *Media, Culture and Society* and *La violencia en la mirada*. Dr. Aran-Ramspott is a member of the International Emmy Awards Jury and the Assessment Committee of the Catalan Public Broadcasting Corporation (CCMA). Her research interests include media representations of violence, social perception of symbolic violence, processes in the influence of television on children and young people, interpersonal relations and social groups as media audiences. Her studies employ qualitative methodology and look at gender identities and ethics, models of love relationships and the media.

Amber L. Ferris is an Assistant Professor of Communication at The University of Akron Wayne College, USA. She received her Ph.D. in Communication Studies from Kent State University, focusing primarily on media effects. She also has a Master's degree in Communication from Michigan State University, and a Bachelor's degree from the same university in English. Dr. Ferris studies the effects of television on viewers, as well as new technologies. Her recent publications have appeared in the journal *Computers in Human Behavior,* focusing on the motivations and personal characteristics of Facebook users regarding self-disclosure online. She also published studies in the *Journal of Communication* on cultivation effects related to reality dating television, as well as altruism on television. Dr. Ferris has published textbooks on media analysis and writing in the communication discipline, as well as book chapters related to mobile communication, free speech zones, and pro-social media content. She has given professional lectures on communication topics to the United States Air Force officers, the Girl Scouts of North East Ohio organization, and the Akron Community Foundation Women's Endowment Fund Forum on Women and Girls. She is an active member of the National Communication Association and the Ohio Communication Association, and serves as current editor for the *Ohio Communication Journal.*

Jo Anna Grant is a Professor in the Department of Communication Studies at California State University, San Bernardino with specializations in Health Communication and Communication and Aging. She earned her Ph.D. at the University of Oklahoma. Most recently, her research has examined media portrayals of health and aging in television series and in images distributed by the Associated Press. She has published in scholarly journals such as *American Communication Journal, Visual Communication Quarterly, Journal of Applied Gerontology, Journal of Cross Cultural Gerontology,* and *Journal of Personal and Social Relationships*, as well as in academic book chapters and textbooks. She is also one of the first professors in the States to develop and teach public speaking courses entirely online. As such, she has consulted major text book publishers and continues to help them develop and improve their online course delivery systems for public speaking and other communication courses.

Emilio S. Hernandez is a graduate of Texas Tech University and a recipient of AT&T Chancellor's Fellowship. He taught courses in interpersonal communication, communication theory, and public speaking and served as a research intern for Accessible Arts Inc - a non-profit organization helping people with disabilities create, explore, and enjoy the arts. His research interests consist of mass media portrayals of romance, computer-mediated communication, and romance at-large.

Amir Hetsroni is a professor of communication at Koç University in Turkey. He published four books and nearly 100 journal articles and book chapters including entries in *Journal of Communication, Sex Roles, Mass*

Communication & Society, Communication Monographs, Social Science Quarterly, and others. A study published in *Journal of Advertising* identified him as one of the 100 most prolific scholars in advertising research. He was Associate Editor of *Communication Research Reports* and editorial board Member in *Journal of Communication, Communication Studies* and *Corporate Communications*. He is also a media celebrity in his home country, Israel, where he takes part in reality shows as consultant and commentator and takes an active role in anti-censorship campaigns.

Heather L. Hundley serves as Professor of Communication Studies at California State University, San Bernardino, USA. She earned her doctorate from the University of Utah. Her scholarship and teaching focus on media studies, critical/cultural studies, representations of identities, gender, and sports. She has published in journals such as *New Media & Society*, *American Communication Journal*, *Communication Quarterly*, *Visual Communication Quarterly*, *Communication Reports*, and *Journal of Men's Studies*. She has co-authored *Views from the Fairway: Media Explorations of Identity in Golf* and co-edited *Examining Identity in Sports Media* (both with Andrew Billings). More recently she published *Mediated Moms: Contemporary Challenges to the Motherhood Myth*, a co-authored book with Sara Hayden. Her contributed chapters are included in *Fallen Sports Heroes, Media, and Celebrity Culture* (edited by Lawrence Wenner) *and The International Encyclopedia of Media Studies: Content and Representation* (edited by Anghaard Valdivia and Sharon Mazzarella).

Colleen Keene completed her PhD in English at La Trobe University, Bundoora Vic., Australia in 1996. She has been published in *Meanjin, Australian Journal of Feminist Studies* and *Metro Magazine*. Dr. Keane worked for many years as a librarian, and currently works as an editor and an occasional writer.

Maria-Jose Masanet holds a PhD in Social Communication, a Bachelor degree in Audiovisual Communication and a Masters degree in Social Communication from Universitat Pompeu Fabra (UPF). She also holds a Masters degree in Training Secondary Education from Universitat Politècnica de Catalunya (UPC). She is postdoctoral researcher at Universitat Pompeu Fabra and a member of the group Audiovisual Communication Research Unit (UNICA). She has participated in different national and international projects such as "Transmedia Literacy. Exploiting transmedia skills and informal learning strategies to improve formal education" financed by the European Union's Horizon 2020 research and innovation programme and "Undergraduate teaching faced with the audiovisual communication competence in a digital environment" financed by the Ministry of Science and Innovation in Spain. She has been a visiting researcher at the Loughborough University (2013), at the Universidad Central de Venezuela (2014), at the Université Sorbonne Nouvelle Paris III (2014) and at the Centre de Liaison de l'Enseignement et des Médias d'Information from Paris (2014). Her research interests include media education, adolescence and youth, series, media representation of sexuality and love relationships, transmedia and gender identities.

Andrea McClanahan, is a professor of communication studies and the coordinator of the women's studies program at East Stroudsburg University of Pennsylvania, USA. She earned her doctorate at Ohio University. Her teaching and research interests reside in the areas of critical representations of gender in the media, communication theory, and public relations. Specifically, she studies alternative representations of femininity and masculinity and alternative life choices such as the choice to remain unmarried or childfree in popular culture. She has published various chapters in edited volumes on TV series like *The Bachelor*, *Big Bang Theory*, and *Teen Mom.*

Pilar Medina-Bravo holds a Ph.D. in Psychology from the University of Barcelona and has been professor at the Autonomous University (UAB) from 1994 to 2009, at Ramon Llull University (URL) from 1996 to 2008, and at Pompeu Fabra University (UPF) since 2008. She is member of the research group 'Communication, Advertising and Society' (CAS) at UPF. She has been the leading researcher in the following studies: "Hey, you, who do you think you are? Gender perspective within couples: Training project for coeducation"; "Gender inequality in romantic relationships: An analysis of the models of symbolic violence in media fiction"; "Symbolic violence and models of love relationships in television series for adolescents and young viewers: Case study ('Porca Miseria')"; "Stereotypes of the world of the couple and its representation in fictional series: Implications for the construction of identity in pre-adolescents and adolescents". Currently, she is the leading researcher of the project "Media representation of the not healthy body image. Development of a tool for children's from 5 to 10 years old" financed by the Ministry of Economy and Competitiveness in Spain.

Narissra Maria Punyanunt-Carter is an associate professor in the department of communication studies at Texas Tech University, USA and a Ph.D. holder from Kent State University. She teaches interpersonal communication, gender, nonverbal, and romantic relationships. Her research interests include romantic relationships, computer-mediated communication, father-daughter communication, and mass media portrayals of romance. She is a consultant for several book publishers including Bedford St. Martin's, Rowan & Littlefield, Oxford, MCGraw-Hill, and Wadsworth. She has published dozens of articles in outlets such as *Southern Communication Journal, Howard Journal of Communication, Communication Research Reports*, and *Communication Quarterly*. She has been advisor for for student organizations and national honour societies and was awarded Advisor of the Year 2013 by the National Society of Collegiate Scholars (NSCS). She is her university's advisor for Delta Phi Omega and Delta Epsilon Psi (the first sorority and fraternity for Asian Americans) and served as a mentor for Mentor Tech (for ten years) at the Teaching, learning, professional and developmental center (TLPDC). She has published nine book chapters, 45 journal articles, one book review and one book. She received a top paper award at a regional conference and was recently elected as Vice Chair Elect of the Mass Communication division of the National Communication Association and was awarded the "Faculty of the Year" title by the Honors College.

Osnat Roth-Cohen is a lecturer and a senior faculty member at the school of Communication at Ariel University, Israel. She teaches also at Bar-Ilan University from which she obtained her Ph.D in Communication. Her pioneering doctoral research presented a new model of the evolution of advertising industries – a topic she has been teaching, researching since 2006. Her interest areas include marketing communication, advertising history, new media and the creative process. She is teaching courses on strategic marketing, marketing communication and advertising agency simulation workshop.

Jennifer Whalen is a doctoral student in the department of communication at the University of South Florida. Her research focuses on relational communication, specifically romantic relationships, but also includes scholarship focusing on personal narrative, health communication, emotion-work and face-maintenance, dialogical communication, and work on ethical communication practices. Her dissertation research focuses on how widow and widower narratives are embodied, projected, and perceived. She instructed and co-instructed various courses, including

Love and Communication, Relationships on Film, Interpersonal Communication, and Public Speaking utilizing critical pedagogy in the classroom. She serve as a departmental liaison for the Graduate Professional Student Council, the Graduate Ambassador's Program, and the Graduate Student Union. I also serves as Faculty Advisor for The Young People's Revolution, T.W.O.L.A., and The Love and Romance Club.

Dorit Zimand-Sheiner is lecturer and senior faculty member at the School of Communication at Ariel University, Israel. She is also a research fellow at the Institute for New Media, Society and politics at same university. Her main research areas are various aspects of product placement and native advertising, new media and creative and the relationship between advertising, consumer society and media. Dr. Zimand-Sheiner doctoral dissertation concerning myths in advertisements in Israeli daily newspapers passed with distinction from Paris 8 University. Her academic career was prefaced by a long career as senior manager in international advertising agencies around Israel.

INDEX

B

C

D

E

F

G

J

K

L

M

N

O

P

Q

R

S

Y